COVERAGE MEASUREMENT IN THE 2010 CENSUS

Panel on Correlation Bias and Coverage Measurement in the 2010 Decennial Census

Robert M. Bell and Michael L. Cohen, *Editors*

Committee on National Statistics
Division of Behavioral and Social Sciences and Education

NATIONAL RESEARCH COUNCIL
OF THE NATIONAL ACADEMIES

THE NATIONAL ACADEMIES PRESS
Washington, D.C.
www.nap.edu

THE NATIONAL ACADEMIES PRESS 500 Fifth Street, N.W. Washington, DC 20001

NOTICE: The project that is the subject of this report was approved by the Governing Board of the National Research Council, whose members are drawn from the councils of the National Academy of Sciences, the National Academy of Engineering, and the Institute of Medicine. The members of the committee responsible for the report were chosen for their special competences and with regard for appropriate balance.

The project that is the subject of this report was supported by contract no. YA1323-04-CN-0006 between the National Academy of Sciences and the U.S. Census Bureau. Support of the work of the Committee on National Statistics is provided by a consortium of federal agencies through a grant from the U.S. National Science Foundation (Number SBR-0112521). Any opinion, findings, conclusions, or recommendations expressed in this publication are those of the author(s) and do not necessarily reflect the views of the organizations or agencies that provided support for the project.

International Standard Book Number-13: 978-0-309-12826-1
International Standard Book Number-10: 0-309-12826-9

Additional copies of this report are available from The National Academies Press, 500 Fifth Street, NW, Lockbox 285, Washington, DC 20055 or (800) 624-6242 or (202) 334-3313 (in the Washington metropolitan area); Internet, http://www.nap.edu.

Printed in the United States of America

Suggested citation: National Research Council (2009). *Coverage Measurement in the 2010 Census*. Panel on Correlation Bias and Coverage Measurement in the 2010 Decennial Census, Robert M. Bell and Michael L. Cohen (Eds.). Committee on National Statistics, Division of Behavioral and Social Sciences and Education. Washington, DC: The National Academies Press.

THE NATIONAL ACADEMIES

Advisers to the Nation on Science, Engineering, and Medicine

The **National Academy of Sciences** is a private, nonprofit, self-perpetuating society of distinguished scholars engaged in scientific and engineering research, dedicated to the furtherance of science and technology and to their use for the general welfare. Upon the authority of the charter granted to it by the Congress in 1863, the Academy has a mandate that requires it to advise the federal government on scientific and technical matters. Dr. Ralph J. Cicerone is president of the National Academy of Sciences.

The **National Academy of Engineering** was established in 1964, under the charter of the National Academy of Sciences, as a parallel organization of outstanding engineers. It is autonomous in its administration and in the selection of its members, sharing with the National Academy of Sciences the responsibility for advising the federal government. The National Academy of Engineering also sponsors engineering programs aimed at meeting national needs, encourages education and research, and recognizes the superior achievements of engineers. Dr. Charles M. Vest is president of the National Academy of Engineering.

The **Institute of Medicine** was established in 1970 by the National Academy of Sciences to secure the services of eminent members of appropriate professions in the examination of policy matters pertaining to the health of the public. The Institute acts under the responsibility given to the National Academy of Sciences by its congressional charter to be an adviser to the federal government and, upon its own initiative, to identify issues of medical care, research, and education. Dr. Harvey V. Fineberg is president of the Institute of Medicine.

The **National Research Council** was organized by the National Academy of Sciences in 1916 to associate the broad community of science and technology with the Academy's purposes of furthering knowledge and advising the federal government. Functioning in accordance with general policies determined by the Academy, the Council has become the principal operating agency of both the National Academy of Sciences and the National Academy of Engineering in providing services to the government, the public, and the scientific and engineering communities. The Council is administered jointly by both Academies and the Institute of Medicine. Dr. Ralph J. Cicerone and Dr. Charles M. Vest are chair and vice chair, respectively, of the National Research Council.

www.national-academies.org

PANEL ON CORRELATION BIAS AND COVERAGE MEASUREMENT IN THE 2010 DECENNIAL CENSUS

ROBERT BELL (*Chair*), AT&T Research Laboratories, Florham Park, New Jersey
LAWRENCE BROWN, Department of Statistics, Wharton School, University of Pennsylvania
RODERICK LITTLE, Departments of Biostatistics and Statistics, University of Michigan
XIAO-LI MENG, Department of Statistics, Harvard University
JEFFREY PASSEL, Pew Hispanic Center, Washington, DC
DONALD YLVISAKER, Department of Statistics, University of California, Los Angeles (emeritus)
ALAN M. ZASLAVSKY, Department of Health Care Policy, Harvard Medical School

MICHAEL L. COHEN, *Study Director*
DANIEL L. CORK, *Senior Program Officer*
AGNES E. GASKIN, *Senior Program Assistant*
BARBARA A. BAILAR, *Consultant*
MEYER ZITTER, *Consultant*

COMMITTEE ON NATIONAL STATISTICS
2007–2008

Acknowledgments

The Panel on Coverage Evaluation and Correlation Bias in the 2010 Census wishes to thank the many people who contributed to our work. The initial idea for the study came from Hermann Habermann, then deputy director of the Census Bureau. Many other Census Bureau personnel were also instrumental in providing assistance. The contracting officer for this study was Philip Gbur, whose efforts should serve as a model of how best to provide for smooth communications between a National Research Council (NRC) panel and its sponsor. Donna Kostanich was extremely generous with her time and that of her staff, all of whom gave excellent summary presentations on the status of their various research efforts. Along with Philip Gbur, Donna Kostanich established a friendly, collegial environment between her staff and the panel.

We thank the staff of the Census Bureau's census coverage and coverage measurement group for their presentations: Tamara Adams, Paul Livermore Auer, William Bell, Pete Davis, Gregg Diffendal, James Farber, Rick Griffin, Tom Mule, Mary Mulry, Sally Obenski, Doug Olson, Robin Pennington, Preston J. Waite, and David Whitford. The Census Bureau also provided on-site access to the A.C.E. Research Database.

Huilin Li of the University of Maryland carried out many difficult computations on this database at the direction of the panel and staff, and we thank her for her patience and expertise. Stephanie Jaros of the University of Washington provided a comprehensive bibliography on ethnography and census undercoverage, and also provided an excellent paper summarizing ethnographic information on intentional reasons for undercoverage in the decennial census.

As consultant to the panel, Barbara Bailar provided important insights on the history of coverage measurement and its implications for 2010. Also, Roger Tourangeau, member of a sister NRC panel on residence rules in the decennial census, assisted the panel in learning about probes for the possibility of alternative residences on both the coverage follow-up interview and the census coverage measurement interview.

The panel is indebted to Eugenia Grohman of the staff of NRC's Division of Behavioral and Social Sciences and Education for her expert technical editing of the draft report. Also, NRC staff Christine Chen, Lance Hunter, and Agnes Gaskin, as always, provided excellent administrative support for the panel.

We are especially grateful to the project's study director, Michael Cohen, who coordinated both the information gathering and report writing processes for the panel. He did a superb job of organizing the panel's often disjointed observations to facilitate creation of this final report. We would also like to thank Dan Cork for helping to organize and oversee the conduct of the meetings of the panel, and for greatly improving the appearance of the panel's reports, and we are extremely grateful to Connie Citro for helping to oversee all aspects of the study from its inception to publication of the final report, asking very perceptive questions during the panel's meetings, rewriting part of the executive summary, and providing enormously useful advice whenever difficult situations arose.

This report has been reviewed in draft form by individuals chosen for their diverse perspectives and technical expertise, in accordance with procedures approved by the NRC's Report Review Committee. The purpose of this independent review is to provide candid and critical comments that will assist the institution in making the published report as sound as possible and to ensure that the report meets institutional standards for objectivity, evidence, and responsiveness to the study charge. The review comments and draft manuscript remain confidential to protect the integrity of the deliberative process.

We thank the following individuals for their participation in the review of this report: Margo Anderson, Department of History, University of Wisconsin; Eugene P. Ericksen, Department of Sociology, Temple University; David McMillen, External Affairs Liaison's Office, National Archives and Records Administration, Washington, DC; Colm A. O'Muircheartaigh, Harris Graduate School of Public Policy Studies, The University of Chicago; Keith Rust, Westat, Inc., Rockville, MD; Herbert L. Smith, Population Studies Center, University of Pennsylvania; and Martin T. Wells, Department of Social Statistics, Cornell University.

Although the reviewers listed above provided many constructive comments and suggestions, they were not asked to endorse the conclusions or recommendations, nor did they see the final draft of the report before its release. The review of the report was overseen by Henry Riecken, professor of behavioral sciences, emeritus, University of Pennsylvania, and John Rolph, Marshall School of Business, University of Southern California. Appointed by the NRC, they were responsible for making certain that an independent examination of the report was carried out in accordance with institutional procedures and that all review comments were carefully considered. Responsibility for the final content of this report, however, rests entirely with the authoring panel and the institution.

Robert M. Bell, *Chair*
Panel on Correlation Bias and Coverage Measurement in the 2010 Decennial Census

Contents

Glossary of Technical Terms

Accuracy and coverage evaluation (A.C.E.). The coverage measurement program based on dual-systems estimation that was used to evaluate the coverage of the 2000 census.

Adjustment. The use of information from coverage measurement programs to modify census counts in an attempt to correct for coverage errors in the census.

Administrative records. Data in administrative files that are used to help administer governmental programs (e.g., to assess eligibility and for funds distribution).

American Community Survey (ACS). An unclustered continuous household survey that collects information similar to that collected on the old decennial census long form. (Estimates were first available in 2006 from information collected in 2005.)

Be Counted. A decennial census coverage improvement program that provides questionnaires in public locations for individuals to fill out and return if they believe they were missed in the census.

Block clusters. Collections of roughly 30 contiguous housing units that the Census Bureau creates for all U.S. households. In urban areas, these are often individual city blocks.

Census coverage measurement (CCM). The coverage measurement program that will be used to evaluate the coverage of the 2010 census; also, the postenumeration survey and other parts of the 2010 coverage measurement program.

Classification and regression trees. A method for fitting either a categorical or continuous response by developing a decision tree that determines subsets of cases whose most frequent responses (for categorical responses) or whose average values (for continuous responses) are used to provide fitted values.

Components of census coverage error. The four possible census errors: omissions, erroneous enumerations, duplications, and enumerations in the wrong location.

Contamination. A situation in which the census processes carried out in the postenumeration block clusters are different from those in the remainder of the country in ways that may affect census counts for those blocks.

Correlation bias. The bias in dual-systems estimation that is due to the correlation of the individual enumeration propensities for the census and those for the postenumeration survey.

Coverage evaluation. The process of developing a quantitative or qualitative assessment of the quality of the counts in a census.

Coverage follow-up interview. A telephone interview in the 2010 census that will follow up those households for which there is information that a coverage error may have occurred. It will also be used for households with more than six members. (This interview in a sense replaces the coverage edit follow-up interview in the 2000 census, which followed up large households and households that had discrepancies between the indicated household size and the number of people for whom individual characteristics were provided.)

Coverage measurement. The process of developing a quantitative assessment of the quality of the counts in a census; hence, a part of coverage evaluation.

Data-defined enumeration. A census enumeration for which two nonimputed characteristics have been collected.

Demographic analysis. An approach to coverage measurement that bridges the counts for a demographic group from one census to the next through the addition of births and immigrants and the subtraction of deaths and emigrants.

Differential undercount. The difference between the net undercount for a particular demographic or geographic domain and the net undercount either for another domain or for the nation. (See also **Net undercount** and **Undercount**.)

Discriminant analysis. A statistical model that uses a set of variables to construct a function that fits a dichotomous response typically by providing probabilities that a case was a member of one of the two groups.

Domain. A collection of individuals defined by various characteristics, usually geographic and demographic.

Dual-systems estimation. An approach to coverage measurement that uses the census and a postenumeration survey as two independent enumerations of a population. The two enumerations are matched to each other to determine how many are identical, with the results input into a statistical model (referred to in other contexts as capture-recapture).

Erroneous enumerations. Individuals enumerated in the decennial census for whom no enumeration should have been made (e.g., short-term visitors).

E-sample. Generally, the sample of enumerations in the decennial census corresponding to households located in the P-sample block clusters.

Fourth cell. The group of individuals in dual-systems estimation that are missed in both the census and in the postenumeration survey.

Geocoding error. Misidentification of the census block in which an address is physically located.

Gross census error. The total number of both undercounting and overcounting errors for a domain.

Imputation. A technique that "fills in" values for nonresponses, usually based on information collected for other, complete data cases.

KE enumerations. In the 2000 census, enumerations that were data defined but considered to be insufficient for purposes of matching in dual-systems estimation.

Logistic regression. A statistical model that uses a logistic function of a linear combination of covariates to estimate the probability that a case was a member of one of two groups.

Master Address File (MAF). The Census Bureau's current collection of addresses for all U.S. residents and businesses. It is used to develop the Decennial Master Address File, which supports the decennial census mailout operation. (See also **Topologically Integrated Geographic Encoding and Referencing System**.)

Net undercount. The difference between the census undercount and census overcount, often expressed as a rate. (See also **Differential undercount** and **Undercount**.)

Nonresponse follow-up. A decennial census operation used to interview households that failed to fill out a census questionnaire.

Omissions. Individuals whom the decennial census failed to enumerate who should have been enumerated.

Postenumeration survey (PES). A national survey that is operationally independent of the decennial census, taken shortly after the census has concluded its data collection, for purposes of coverage evaluation based on dual-systems estimation.

Poststratification. The use of covariates to define what are believed to be more homogeneous subgroups of the population for which separate dual-systems estimation computations are conducted.

Proxy enumeration. Information collected from landlords or neighbors in the place of information that was intended to be collected from a household's residents.

P-sample. Generally, the residents of the set of households located in the sample of block clusters selected to be included in the postenumeration survey to support dual-systems estimation.

Statistical Administrative Records System (StARS). A national roster constructed for research purposes by the Census Bureau that links people to current addresses by merging and unduplicating a number of administrative records.

Synthetic estimation. A statistical method for small-area estimation that assumes that net undercount rates for demographic groups in a domain apply without change to all geographic subsets of that domain.

Topologically Integrated Geographic Encoding and Referencing System (TIGER). A geographic information system that links a given address to a physical location defined by city blocks, roads and railroads, natural boundaries, and political boundaries.

Undercount. The measurement of either the number or rate of individuals missed in the census that should have been enumerated. (See also **Differential undercount** and **Net undercount**.)

Executive Summary

The U.S. Census Bureau is justifiably proud of its more than 50-year history of evaluating the degree of net coverage error (undercoverage minus overcoverage) of the population in the decennial census. In addition to the information provided by the coverage measurement programs, this effort has resulted in the development of two internationally used census coverage measurement methods, dual-systems estimation and demographic analysis. Dual-systems estimation uses a coverage measurement post-enumeration survey as an independent enumeration of a population. Based on the number of matches between the census and the post-enumeration survey enumerations, an estimate of the number missed by both enumerations is generated, and therefore an estimate of the size of the population. Demographic analysis uses an accounting relation to estimate a population group's size, adding the result of births and immigration, and subtracting those as the result of deaths and emigration.

The census coverage measurement programs have historically addressed three primary objectives with varying degrees of emphasis: (1) to inform users about the quality of the census counts for various applications; (2) to help identify sources of error to improve census taking; and (3) to provide alternative ("adjusted") counts based on information from the coverage measurement program. In planning the 1990 and 2000 censuses, the main objective was to produce alternative counts based on the measurement of net coverage error, although the alternative counts were never used for either reapportionment or redistricting. Subsequently, a 1999 Supreme Court decision precluded the use of alternative counts, when based on sampling, for use in reapportionment. In addition, it is difficult to provide alternative counts in time for reapportioning congressional districts. Consequently, for the 2010 census coverage measurement program, the Census Bureau has stated its intent to deemphasize the goal of providing alternative counts and is instead planning on focusing its coverage measurement program on the second goal of improving census processes.

The panel strongly supports the Census Bureau's change in goal. However, the panel finds that the current plans for data collection, data analysis, and data products are still too oriented toward measurement of net coverage error to fully exploit this new focus. Although the Census Bureau has taken several important steps to revise data collection and analysis procedures and data products, the panel recommends further steps to enhance the value of coverage measurement for the improvement of future census processes.

Recommendation 1: The Census Bureau should more completely shift its focus in coverage measurement from that of collecting data and developing statistical models with the goal of estimating net coverage error to that of collecting data and developing statistical models that support the improvement of census processes.

To help achieve this new goal, instead of only measuring net census error, the Census Bureau also plans to measure the four components of census coverage error: (1) census omissions, (2) census duplications, (3) erroneous census enumerations, and (4) census enumerations in the wrong location. The panel supports these plans, since different types of coverage errors are caused by different interactions between census processes and housing units and their occupants. The estimation of these four components of coverage error can be supported by the general structure of the data collection and matching that is carried out in support of dual-systems estimation, though modified and expanded to support this different purpose. The panel finds, however, that the Bureau's plans could be more fully developed for this purpose.

Recommendation 9: The Census Bureau should further develop and refine its framework for defining the four basic types of census coverage error and measuring their frequency of occurrence. The Census Bureau should also develop plans for operationalizing the measurement of these components using data from the census and the census coverage measurement program.

Certain types of housing units are more likely to be missed than others, yet the Bureau's current design for the coverage measurement postenumeration survey does not adequately take this into account.

Recommendation 6: The Census Bureau should compare its sample design for the 2010 census coverage measurement postenumeration survey with alternate designs that give greater sampling probability to housing units that are anticipated to be hard to enumerate. If an alternate design proves preferable for the joint goals of estimating component coverage error and net coverage error estimation, such a design should be used in place of the current sample design.

Thorough analysis of data from the coverage measurement survey offers a unique opportunity to learn how census errors occur and how census processes might be changed to reduce them in the future. Working with outside researchers to the extent possible, the Census Bureau should study and give consideration to a richer menu of analytic methods using data collected from the coverage measurement postenumeration survey.

To date, the Census Bureau has not given sufficient attention to developing statistical models that link the frequency of the four components of coverage error to census processes, person and housing characteristics, and other predictors. These models, which can be thought of as forms of discriminant analysis, could use a wide variety of approaches, including logistic regression and various data mining methods, such as classification trees, support vector machines, and neural nets. It may be that modeling the frequency of erroneous enumerations may benefit from an entirely different approach than modeling the frequency of census duplicate enumerations, or census omissions. Consideration should also be given to the potential for using predictor variables that are specific to each type of error. Also, the use of separate models for distinct population subgroups should be considered.

Recommendation 12: The Census Bureau should develop regression models that elucidate the various types of census coverage error, using specified dependent and predictor variables. To the extent that the database supporting these models can be made available to external researchers, it is extremely important that the Census Bureau pursue all viable avenues to involve outside researchers in the development of such models.

Recommendation 10: In developing the logistic regression models or other types of discriminant-analysis models of match status,

correct enumeration status, and components of census coverage error, the Census Bureau should consider:

- **Use of several approaches before focusing on a specific model; besides logistic regression, alternatives should include use of other link functions, discriminant analysis, and various data mining approaches, such as classification trees, support vector machines, and neural nets.**
- **Thorough examination of the subset of predictors that is best suited to each individual statistical model; the predictors for these various statistical models need not be identical; however, there may be a benefit to constraining the (logistic regression) models of match rate and correct enumeration rate to have identical variables in the estimation of net coverage error, and research should be carried out to assess whether this benefit outweighs the benefit of selecting variables that are optimal for each of these two logistic regression models.**
- **To effectively blend information from auxiliary sources at various levels of geographic and demographic aggregation, random effects modeling and Bayes' methods also should be examined.**

This effort will require that considerable resources be allocated for the development and use of these models, comprising essentially a new Census Bureau research program.

Recommendation 2: The Census Bureau should allocate sufficient resources, including funding and staff, to assemble and support an ongoing intercensal research program on decennial census improvement. Such a group should focus on using the data from the census and the census coverage measurement programs to identify deficient census processes and to propose better alternatives. The work of this group should be used to help design the census tests early in the next decade.

To do this, it is important that sufficient data from 2010 be retained both for the measurement of the components of census coverage error and to provide the predictors that might be useful in these models.

Recommendation 13: For a sample of households, the Census Bureau should retain data that provide a comprehensive picture of the census processes used to enumerate it, and the individuals residing in it, to facilitate subsequent evaluation. To allow linking assessment of census coverage error with a history of the census

processes, this sample should substantially overlap with the census coverage measurement sample.

The creation and exploitation of an analytic database, in order to improve census processes, should be the primary goal of the coverage measurement program in 2010. However, the Census Bureau is focusing instead on producing summary tabulations related to the frequency of the components of coverage error by major census process. Such tabulations will have little value: the complexity of coverage error requires a more sophisticated use of the data through development of statistical models for each component of coverage error.

Recommendation 11: The primary output of the Census Bureau's coverage measurement program in 2010 should be an analytic database that is used to support the development of statistical models to inform census process improvement. The production of summary tabulations should be of lesser priority.

In addition to the topics discussed here, the panel also offers recommendations in five areas: (1) the need to retain comprehensive information on the functioning of the coverage follow-up interview, (2) the timing of the coverage follow-up interview in relationship to the timing of the census coverage measurement data collection, (3) the testing of administrative records for various census purposes, (4) the development of improved techniques for treating missing data in coverage measurement models, and (5) research to guide improvement of demographic analysis.

1

Introduction

The goal of the decennial census is to count everyone in the country, once and in the right place, for the purpose of allocating representation in Congress. The census satisfies this goal only incompletely, as some people are omitted that should be included, and some enumerations are either duplicates, are in the wrong location, or are either not residents of the United States or are not people. These four components of coverage error have an important impact on the representation of demographic groups and geographic jurisdictions in Congress.

PROGRAM OBJECTIVES

Since the 1950 census there has been an effort by the Census Bureau to estimate the size of error in census counts for areas and demographic groups and to use the information to improve census processes. The programs to measure census coverage error are referred to as coverage measurement programs. In recent years, coverage measurement programs included a third objective—correcting the census for enumeration error, referred to as census adjustment. The techniques used in coverage measurement programs to understand the extent of enumeration errors are sample surveys, dual-systems estimation (DSE), and demographic analysis.

In contrast to the previous two censuses, the Census Bureau has decided that the 2010 census coverage measurement (CCM) program will have a new principal objective: to emphasize census improvement rather than census correction.[1] As a result of this change, rather than focus on the measurement of net census coverage error for demographic and geographic subsets of the U.S. population, the coverage measurement program in 2010 will focus on measurement of the rates of the components of census coverage error for subsets of the population defined not only geographically and demographically, but also by the census processes used. The hope is to use this information to help identify census processes that are associated with a high rate of coverage error and then identify alternative processes to reduce the rates. This feedback loop will then help to facilitate census process improvement for subsequent censuses. Of course, an important secondary goal of the coverage measurement program still remains, which is to inform census data users about the net coverage error for large geographic areas and demographic groups. The shift in the principal objective of the coverage measurement program from that adopted in 2000 (and in 1990) stems from both the specific circumstances surrounding the 2000 census and broader dynamics. With respect to the 2000 census, a decision by the Supreme Court in 1999 precluded the use of census adjustment for purposes of apportionment of the U.S. House of Representatives if it is based on data from a sample survey (as it would almost certainly be). Furthermore, the time needed to carry out and review coverage measurement also very likely precludes the use of adjusted counts as input into redrawing the boundaries of the districts for the U.S. House of Representatives (unless the dates for the census, April 1, or for redistricting, April 1 of the following year, are changed).

Also, the problem of census omissions has become a problem of erroneous enumerations (overcounts) and census omissions: Prior to 1990 the main coverage problem was census omissions, but at the national level in 2000 the number of erroneous enumerations was roughly the same as the number of census omissions.[2]

The new problem of both census omissions and erroneous enumerations has arisen partly because of the effort to reduce the main problem of census omissions that dominated prior to 2000. In response to the challenge of reducing census omissions, between 1960 and 2000 the Census Bureau added a number of alternative ways in which households could be included on the Master Address File (including the Local Update of Census Addresses Program), and in which individuals could be enumerated in the census (including the Be Counted Program). These additional ways for households and individuals to be included in the census certainly

[1]Actually, this is in a sense a return to the pre-1990 goal of coverage measurement.

[2]While this balancing did not obtain for every demographic or geographic subgroup, it is also true that the differential nature of net coverage error was reduced from that of previous censuses. (For information on adjusted counts in 2000, see Schindler, 2006.)

increased the number of duplicate enumerations, which contributed to the "balancing" of the undercount and the overcount in the 2000 census.

In addition to the duplication resulting from new avenues for enumeration, there is evidence that a number of social dynamics are also increasing the potential for census overcounts (see National Research Council, 2006). First, the structure of households is becoming more complicated, with more people having attachments to multiple households, including children in shared custody. Second, the number of people with multiple residences is increasing: This group includes people with vacation homes and "commuter marriages." It has also been hypothesized that the quality of the enumerator workforce has decreased over time.

The shift in the principal objectives of coverage measurement raises many interesting and important technical issues. For example: What sample design for the coverage measurement survey should be used? What estimation approaches should be used in support of the attempt to link error status to relevant census processes? What data products would best communicate the linkages between census component coverage error and census processes in need of improvement?

PANEL CHARGE AND WORK PLAN

At the Census Bureau's request, the National Academies established the Panel on Correlation Bias and Coverage Measurement in the 2010 Decennial Census to examine the Census Bureau's coverage measurement plans for 2010 with the following charge:

> This project involves a study of four issues concerning census coverage estimation with the goal of developing improved methods for use in evaluating coverage of the 2010 census. A panel of experts will conduct the study under the auspices of the Committee on National Statistics of the Division of Behavioral and Social Sciences and Education. The panel is charged to review Census Bureau work on these topics and recommend directions for research. The panel's work may require development of statistical models to extend the dual-systems estimation (DSE) approach, and may also include suggestions for the use of auxiliary data sources such as administrative records. DSE, as applied to the 1990 and 2000 censuses, had several benefits as well as limitations as a means for estimating net census coverage. Some of the limitations were:
>
> 1. The approach was designed for estimating net census coverage errors and did not provide accurate estimates of gross coverage errors, i.e., of gross census omissions separate from gross census erroneous enumerations. In the DSE approach applied in the 1990 and 2000 censuses, certain census enumerations classified as erroneous were balanced against certain coverage survey cases classified as nonmatches (census

omissions) for the purpose of estimating net census coverage. Some of these paired census enumerations and coverage survey cases did not necessarily reflect gross errors.

2. The application of DSE in Accuracy and Coverage Evaluation (A.C.E.) Revision II during the 2000 census accounted for duplicates found in the census in a simplistic way due to lack of information as to which member of a duplicate pair was a correct enumeration and which was an erroneous enumeration. This led to estimation error, as did the simplistic treatment of A.C.E. cases (P-sample) that matched to census enumerations outside the search area.

3. The poststratification approach used to apply the DSE had certain limitations. First, the number of factors that could be included in the poststratification was limited because the approach cross-classified the factors, so that each factor added to the poststratification greatly split the sample. (Collapsing of poststrata was needed because many of the cross-classified cells had small sample sizes.) Second, the synthetic error that arose from the synthetic application of the poststratum coverage correction factors to produce estimates for subnational areas and population subgroups was not reflected in their corresponding variance estimates.

4. Comparisons of aggregate tabulations of DSEs with estimates from demographic analysis (DA), in both 1990 and 2000, suggested underestimation by DSE of persons missed by both the census and the coverage survey (correlation bias). In the 2000 A.C.E. Revision II, sex ratios from DA were used to determine factors to correct adult male estimates for correlation bias, assuming no correlation bias for children and adult females. This approach appeared effective for adult blacks, but there were concerns about the appropriateness of its assumptions for other race/origin groups (particularly Hispanics). Also, DA totals for young children (0–9) exceeded the corresponding aggregated DSEs from A.C.E. Revision II by a sufficient amount to suggest possible correlation bias in estimates for young children.

The Census Bureau is interested in improving the DSE methodology to address the above issues to the extent possible, to develop improved methods for estimating coverage of the 2010 census, both in regard to net errors and gross errors.

This original charge to the panel had four areas of focus: (1) to effectively measure the components of coverage error rather than net coverage error; (2) to improve the determination of duplicate status and the measurement of the rate of census duplication; (3) to assess the use of model-based alternatives to poststratification, including their impact on the ability to model local heterogeneous effects; and (4) to examine the use of demographic analysis to correct for correlation bias. It was also understood that the panel's work might involve the review of other statistical models proposed for estimation of net coverage error and the use of auxiliary data sources, such as administrative records, in DSE. Consistent with this, it was recognized that all the data retained from the 2010 census—not only the census enumerations themselves and the postenumeration survey and matching results, but also data collected by the various management information systems that monitor census processes—could prove useful in modeling census error rates and providing information on the sources of census error. Therefore, the panel was also asked to provide advice on what data should be retained from the 2010 census.

During the course of the study, several other issues in connection with the panel's overall task arose: a review of the Census Bureau's draft document providing a framework for defining and estimating components of census coverage error; examination of the possibility of estimating the match status of cases previously categorized as having insufficient information for matching, in order to reduce the number of cases classified as erroneous enumerations due to item nonresponse; assessment of the various alternatives that could be used to reduce or address contamination due to the similarity and simultaneity of the census coverage follow-up interviews and the initial CCM interviews in 2010; the CCM postenumeration survey design. More generally, the panel considered any other limitations that the 2000 A.C.E. Program had in addressing the objective in 2010 of measuring the rate of census component coverage error.

The panel took as given the basic design of data collection and matching operations planned for census coverage measurement in 2010. The plans include a sizable postenumeration survey that will be matched to the census to assess match status for the housing units (and individual residents in those housing units) found in a sample of census block clusters. The panel examined modifiable aspects of the data collection for the 2010 coverage measurement program, including the sample design, seeking possible improvements. The panel did *not* address the broader range of possible coverage measurement programs that might best support census improvement over time.

A postenumeration survey that is matched to the census, along with a sample of census records that are matched to the census enumerations, can be used to directly support the new objective of census improvement because one can identify individual census enumerations that are duplicates, erroneous enumerations, and enumerations in the wrong location. Furthermore, one can identify a sample of individuals that were omitted in the census enumerations. In addition, and crucially, one can identify the census processes that were used to enumerate these individuals, along with characteristics of the individuals, their households and housing units, and contextual variables. This information can then be analyzed using statistical models to link higher rates of each of the four types of census error and the associated census processes. Thus, a data collection and estimation program that was originally proposed to be used in an aggregate way for estimating net coverage error for large demographic and geographic groups is also very useful for identifying individuals of interest to populate a database to support statistical models predicting census coverage error. The change in objectives also suggests that rather than try to "fix" the census for net undercoverage using sampling-based statistical procedures, it may be preferable to use information on census coverage error to identify deficiencies in the decennial census processes.

Finally, in the course of its work, the panel also considered the possible benefits of a broader program of research on census coverage measurement. The panel explored other activities that might support measurement of components of census coverage error. This work was undertaken while recognizing that plans are close to final as the 2010 census nears, with a view to plans for coverage measurement for 2020.

In sum, the panel undertook to evaluate the Census Bureau's plans for coverage measurement in the 2010 census and to provide suggestions and recommendations for changes and additions to those plans, given the new objective of measuring the rates of components of census coverage error, with the ultimate goal of assessing the contribution of various census component processes to census coverage error.

PLAN OF THE REPORT

Substantial portions of this report are taken from the material in the panel's interim report (National Research Council, 2007). This report expands the panel's work in five areas: assessment of duplicate status, missing data methods, the census coverage measurement sample design, improvements to demographic analysis, and treatment of the potential contamination of the census coverage measurement sample interview by the overlap in the field with the census coverage follow-up interview.

To collect the necessary information for this study, the panel held six plenary meetings between August 2004 and July 2007. During the course of our meetings, Census Bureau staff described their current coverage measurement research activities and intended directions for further work, their test and dress rehearsal plans, and their plans for the 2010 CCM program.

Some of the Census Bureau's research on net coverage error has been facilitated by the development of an A.C.E. research database. This database contains the data collected by A.C.E. to support estimation of net coverage error in 2000, and it is weighted to represent the additional information collected from the national duplicates search and the evaluation follow-up survey so that the net coverage error estimates produced are nearly identical to those from A.C.E. Revision II.

This introductory chapter is followed by four chapters and three appendices. Chapter 2 first discusses types of census coverage error and the coverage error metrics for domains of interest. It then describes the three primary purposes for coverage measurement and DSE and demographic analysis, the two primary methods used to measure net coverage error. Chapter 2 also presents short histories of the U.S. census coverage measurement programs from 1950 to 1990, including a description of A.C.E., the coverage measurement program for the 2000 census.

Chapter 3 examines how the 2010 census differs from the 2000 census with respect to the impact on the coverage measurement program for 2010. It looks in some depth at the treatment of duplicates in the 2010 census and the 2010 coverage measurement program, including the possibility of contamination of the 2010 coverage measurement data collection through the application of the coverage follow-up interview. The chapter also discusses how the use of administrative records could potentially assist in both coverage improvement and coverage measurement for the 2010 census.

Chapter 4 discusses a number of technical topics introduced by the various changes made in coverage measurement for 2010, including: the sample design for the census coverage measurement postenumeration survey in 2010; the use of logistic regression modeling as a substitute for poststratification in modeling net coverage error; how one compares competing models in this situation; and the treatment of missing data in net coverage error modeling, including the Census Bureau's current plans for addressing missing data prior to fitting the logistic regression models in 2010. In relation to the issue of missing data, the chapter includes a description of an attempt by the Census Bureau to greatly reduce the number of cases that are considered to have insufficient information to support matching. The chapter concludes with a discussion of how to improve demographic analysis for use in census coverage measurement in 2010.

Chapter 5 first briefly outlines the Census Bureau's framework for defining and estimating components of census coverage error. It then considers potential variables for use in statistical models to assess correlates of components of census coverage error. The chapter ends with a consideration of the purpose of the key output from the census coverage measurement program in 2010—the analytic capability to develop statistical models linking census coverage errors of various types to individual and household characteristics and census process variables.

There are three appendixes. Appendix A provides additional details from the paper by Mulry and Kostanich (2006). Appendix B provides additional details on the use of logistic regression models as a substitute for poststratification. Appendix C provides biographical sketches of panel members and staff.

2

Fundamentals of Coverage Measurement

The decennial census is used for a wide variety of purposes by federal, state, and local governments, by businesses, and by academe. However, the Constitutional goal of the census is to allocate the population to the states and local areas to support apportionment and congressional redistricting. This use of the census counts makes determination of the correct location of enumerations especially important and also focuses attention on racial differentials. Further, due to racial segregation, differential net undercoverage is likely to impact geographic differential undercoverage. Clearly, the broad goal of measuring the quality of the coverage of the census is to assess the extent of census coverage error by domain and by demographic group.

Coverage measurement is a collection of techniques that measure the differences between census enumerations and the corresponding true counts for groups or areas. Coverage measurement is the quantitative aspect of coverage evaluation, which also encompasses more qualitative techniques, such as ethnographic observation. The differences between census counts and the corresponding true counts at the level of the individual (or the household) are referred to collectively as census coverage errors, and in this chapter we categorize types of census coverage error and indicate methods that can be used for their summarization. We then detail the three primary (potential) uses of census coverage measurement that rely on summarizations. Finally, we provide a brief overview of the methods that are currently used in the U.S. census for coverage measurement.

TYPES OF CENSUS ERRORS

There are two obvious ways in which the census count for an individual can be in error: A person could be included in the census as an enumeration when he or she should have been omitted—an overcount—or the person could be omitted from the census when she or he should have been included—an undercount. In addition, since the primary applications of census counts are for apportionment of the states and redrawing of congressional districts, it is important that each individual be counted in their appropriate location. When a person is counted in other than the correct location, the effect of this error depends on both the distance between the recorded location and the true location and on the intended application of the counts (see below).

Given that, we decided in this report to separately categorize undercounts from overcounts, which are always errors regardless of the location of the enumeration, and those from enumeration errors that result from counts in the wrong location. This approach is not due to any sense that the latter errors are less important, but that they have different causes and therefore different solutions, and second that they are of different types as a result of the various degrees of displacement.

This classification of census coverage error differs from the classification that has been typical up now. In that classification scheme, an overcount was any erroneously included enumeration, which included enumerations that were in the wrong location, regardless of whether the error was a few blocks or hundreds of miles. Similarly, an undercount was any erroneously omitted enumeration, which included enumerations that were in the census but were attributed to another (incorrect) location. As a result, in the previous scheme, an enumeration in the wrong location was represented as two errors: an overcount for the location that was recorded and an undercount at the correct location. The approach adopted here for classifying coverage error is consistent with a framework developed by Mulry and Kostanich (2006), which is described in Chapter 5. We now provide more detail on the nature and causes of these various types of census coverage error.

Undercounts

Omissions result from a missed address on the decennial census' Master Address File (MAF), a missed housing unit in a multiunit residence in which other residences were enumerated, a missed individual in a household with other enumerated people, or people missed due to having no usual residence.

Overcounts

Overcounts result from including enumerations that should not have been included in the census and from counting people more than once. Enumerations that should not have been included in the census are for people who were not residents of the United States on Census Day, and includes those born after census day and those who died prior to census day; people in the United States temporarily; and enumerations of fictitious people. As explained above, we restrict the term "erroneous enumerations" to those enumerations that should not have been included in the census anywhere at all, thereby excluding duplicates and those counted in the wrong location.

Duplicates

Duplicates can result from: (1) repeat enumerations of a subset of the individuals from a household, sometimes as a result of the multiple opportunities for being enumerated in the census; (2) an address being represented in more than one way on the MAF, resulting in the duplication of all residents; and (3) the inclusion of a person at two distinct residences, possibly both of which are part-time residences or because of a move shortly before or shortly after Census Day.

Counts in the Wrong Location

The two fundamental types of census coverage error, overcounts and omissions (undercounts), reduce the accuracy of the total count for the people in any geographic area that contains or should contain the individual counted in error. In addition, as mentioned above, there can also be errors in the geographic location of an individual or an entire household, which can also impact the accuracy of census counts.

Counting a person in the wrong location can result from a misunderstanding of the census residence rules and the resulting reporting of someone in the wrong residence. This can result from having an enumerator assign a person to the wrong choice from among several part-time residences or from the Census Bureau's placing an address in the wrong census geographic location (called a geocoding error). Placing a person in the wrong geographic area will lower the count for the correct geographic area and raise it for the incorrectly designated area. Therefore, whether there is an effect on census accuracy depends on the distance between the correct and incorrect locations and on the summary tabulation in question: the more detailed the tabulation is with respect to geography, or the greater the displacement, the greater the chance that geographic errors will affect the quality of the associated counts. Placing a person in the wrong location can therefore result in zero additional errors or two additional errors. (One additional error is also possible, by placing a duplicate enumeration in the wrong location.)

Demographic Errors

A similar outcome will occur when a person's demographic characteristics are recorded in error. This happens when a person is assigned to the wrong demographic group through a reporting error or through use of imputation of an individual's demographic characteristics when those characteristics are not provided by the respondent. Again, placing a person in the wrong demographic group will lower the count for the correct demographic group and raise it for the incorrectly designated demographic group. Whether this error has an effect on the decennial census counts depends on the aggregate of interest: as above, the more detailed the tabulation demographically, the greater the chance that demographic errors will affect the quality of the associated counts.

Imputations

In addition to census coverage errors that result from the data collected in the census, there are also enumeration errors that result from the methods, typically imputation, that are used to address census nonresponse. As mentioned in National Research Council (2004a), in addition to item imputation (which is used to address missing characteristics for so-called data-defined enumerations), there are five different degrees of "missingness" for the residents of a housing unit that can result in five different types of whole person or whole household imputation (count imputation). Imputation methods used to address whole household nonresponse will often result in counts for a housing unit that do not agree with the true number of residents of that housing unit, and these differences contribute to coverage error. However, we assert that the discrepancies that result from the application of an imputation technique are not errors of either omission or overcoverage and therefore should not contribute to assessments of the magnitudes of the components of census coverage error. Whole household imputation is simply a means for producing counts that are as accurate as possible when aggregated for various domains of interest.[1] Thus, the effectiveness of an imputation algorithm should be assessed by its aggregate performance (e.g., bias, variance, mean-square error for domains of interest) and should not be

---

[1]We use the term "domain" to refer to any demographic or geographic aggregate of interest.

considered as correct or incorrect at the level of the household. To sum up then, there are four basic types or components of census coverage error: omissions, duplicates, erroneous enumerations, and enumerations in the wrong location.

COVERAGE ERROR METRICS FOR AGGREGATES

Since census coverage errors can be positive (overcounts) or negative (undercounts), they can partially cancel each other out when census counts are aggregated over a domain. Specifically, the difference between the census count and the true count for a domain is equal to the number of overcounts minus the number of undercounts, plus the net from enumerations in the wrong location for the residents of the housing units in that domain. The *net coverage error* or the *net undercount*, defined as the difference between the census count and the true count for a domain, is therefore a useful assessment of the effect of census coverage error on an aggregate of interest.

Net coverage error has two benefits: (1) it directly assesses the utility of census counts for aggregates of interest, and (2) it can be compared with previously published estimates of net coverage error for historical comparisons of census quality. *Percent net undercount* expresses net coverage error as a percentage of the true count and therefore facilitates comparison of the net coverage error between domains. *Differential net undercount*, the difference between the percentage net undercount for a specific domain and the percentage net undercount for another domain (or for the nation), is therefore a useful measure of the degree to which one domain is (net) undercounted relative to another. To be precise, let C_i be the census count for the ith domain, and let C_+ be the census national total. Similarly, let T_i and T_+ be, respectively, the true count for the ith domain and the true national total. Then the differential net undercount is

$$\frac{C_i - T_i}{T_i} - \frac{C_+ - T_+}{T_+} = \frac{C_i}{T_i} - \frac{C_+}{T_+}.$$

Many uses of census data (e.g., apportionment and fund allocation) depend on census counts as proportional shares of the population, rather than as population counts, and in those situations a measure of the quality of the counts for a domain of interest is

$$\frac{C_i}{C_+} - \frac{T_i}{T_+}.$$

For comparison of the quality of two sets of estimated counts used as counts, a common yardstick is the sum of squared net errors over domains. When comparing the quality of two sets of estimated counts used as population shares, the error of the shares is again commonly summarized, with errors as the difference between population shares and true shares, by adding squared errors over domains, but now *weighted by the population size* (T_i), since otherwise one is equating a given error in population shares for a small and a large domain. Specifically, the following loss function would be reasonable to use:

$$\sum_i \left(\frac{C_i}{C_+} - \frac{T_i}{T_+} \right)^2 T_i .$$

Although it is clearly very useful, net census error, or net undercount, is an inappropriate summary assessment of census coverage error when the objective is census improvement because a substantial number of overcounts and undercounts may cancel each other for a given domain, which may obscure problems with census processes. Also, while these errors may balance each other for a given domain for a given census, they may not balance to the same extent either in more detailed aggregates or in subsequent censuses.

To address this possible imbalance, some have argued for tabulating census *gross error*, which is the sum of the number of errors, overcounts, undercounts, and errors in the wrong location, relative to domains of interest. However, there are two problems with gross error as a summary measure of the quality of the census enumeration process. First, as noted above, enumerations in the wrong location will only matter when the degree of displacement and the tabulation in question are such that the displacement places someone in the wrong tabulation cell. Therefore, enumerations in the wrong location should not be interpreted as equivalent to overcounts or omissions. Furthermore, census coverage errors, which we classify as erroneous enumerations, duplicates, omissions, and counts in the wrong place, all have somewhat different causes. Given the current objective of supporting a feedback loop for census improvement, it is important to separate out the summaries of these various components so that their magnitudes can be assessed individually, rather than trying to aggregate them into a single error measure. Second, for counts in the wrong location, rather than a percentage error measure—which is an appropriate summary measure for omissions, erroneous enumerations, and duplications—a more useful summary assessment would provide a representation of the frequency of enumerations in the wrong location as a function of some representation of the degree of the displacement (so that location error rates would diminish as the displacement increases). This approach would facilitate the assessment of the degree to which errors from enumerations in the wrong place effect various applications of the counts.

The term *components of (census coverage) error* communicates this idea of separating out the enumeration errors into these categories of duplications, erroneous enumerations, omissions, and geographic errors so that their individual causes can be better analyzed.

For sake of completeness, we again mention that there are also errors in counts that are attributable to errors in a person's demographic characteristics, and there can, of course, also be errors in a person's other characteristics, for example, whether the residents own or rent their housing units. These are ignored in this discussion, though errors in characteristics used to model net coverage error can negatively affect its estimation, and it is therefore important to reduce the frequency of such errors.

Whether one uses net coverage error or rates of components of census coverage error to represent the quality of the census counts for a domain clearly depends on the analysis that one has in mind. To support as much flexibility in summarization and analysis as possible, information on census coverage error needs to be retained at as basic a level as possible, in addition to the summary tabulations that the Census Bureau provides. This retention would have two advantages. First, it would permit a more precise assessment of the effect of census errors on any specific application of the counts. For instance, one could assess the impact of omissions (ignoring the extent to which they are offset by overcoverage errors) on a specific domain of interest that is not provided in the standard Census Bureau tabulations from the coverage measurement program. Second, and more importantly, retention of information on census coverage error at the level of the individual allows for the examination of (causal) associations using statistical models that relate whether a coverage error was or was not made as a function of the census enumeration processes used and individual and housing unit characteristics. Such an analysis could also include correlates of whole-household omissions, correlates of omission errors that only affected some residents of a household, correlates of whole-household duplications, correlates of partial-household duplications, or correlates of the coverage error of counting individuals in the wrong place (for various degrees of misplacement).

In sum, there are various components of census error that have various applications, and there is therefore a need for access to those errors at an individual level and to link those errors to potential causal factors to support various descriptive and analytic needs. By "descriptive" we mean summary assessments of the quality of census counts for domains; by "analytic" we mean the development of statistical models that attempt to discriminate between individuals and households that are and are not counted in error.

PURPOSES

Coverage measurement has historically served multiple purposes. Since its earliest inception in the 1950 census, it has had the goal of evaluating the accuracy of census counts for geographic and demographic domains, with a focus on assessing net error for domains. The primary goal was to inform users as to the quality of the census counts for various applications. In addition, but to a much lesser extent, coverage measurement has also been used to provide information relevant to developing a better understanding of census process inadequacies, leading to improvements in design for the subsequent census.

The estimation of net error has also raised the possibility of providing alternative counts for use in formal applications, known as census adjustment. We know of only one formal use of adjusted census counts to date, namely, the use of adjusted counts to modify population controls used for the Current Population Survey (CPS), the National Health Interview Survey, the National Crime Victimization Survey, and the Survey of Income and Program Participation during the 1990s, which in turn affected the estimate of the number of people unemployed during the 1990–2000 intercensal period by the Bureau of Labor Statistics. However, the primary focus of coverage measurement in both 1990 and in 2000 was to produce adjusted census counts for official purposes, assuming that it could be demonstrated that the adjusted counts would be preferred to the unadjusted census counts for apportionment and redistricting.

The stated Census Bureau plan that the primary purpose of the coverage measurement program in 2010 would be to measure the components of census coverage error in order to initiate a feedback loop for census process improvement is a substantial innovation. An interesting question is the extent to which a coverage measurement program can be used for this purpose, and a major charge to this panel was to determine the extent to which this new focus of coverage measurement should affect the design of the coverage measurement program and the resulting output and analyses.

Evaluation of the Accuracy of the Census Counts

Census counts serve a variety of important purposes for the nation, including apportionment, legislative redistricting, fund allocation, governmental planning, and support of many private uses, such as business planning. Users of census data need to know how accurate the counts are in order to determine how well they can support these various applications. The needed information includes an understanding of the extent to which the accuracy of census counts differ by location or by demographic group and the extent to which accuracy has improved from one census to the next.

The total population count of the United States is probably the most visible output of a census, so one obvious measure of coverage accuracy for the census is the error in the count for the entire United States over all demographic groups. However, essentially all applications of the census—e.g., redistricting and local planning—use population counts at various levels of geographic and demographic detail. Consequently, it is important to assess the rates of net undercoverage by various geographic or demographic domains.

Historically, a key issue has been, and remains, the differential net undercount of blacks, Hispanics, and Native Americans, which has resulted in the repeated underrepresentation of areas in which those groups make up a large fraction of the residents. In particular, the differential net undercount of these groups has led to their receiving less than their share of federal funds and political representation (see, e.g., Ericksen et al., 1991, for more details). Given this, it is as important as ever for the Census Bureau, in evaluating possible alternative designs for the decennial census, to not only assess the likely impacts on the frequency of components of census coverage error, but also to assess the impacts on differential net coverage error for historically undercounted minority groups.

Census Adjustment

The 1999 Supreme Court decision (*Department of Commerce v. United States House of Representatives,* 525 U.S. 316) precluded the use of adjustment based on a sample survey for congressional apportionment. In addition, the Census Bureau concluded that time constraints currently preclude the computation and evaluation of adjusted counts (based on a postenumeration survey) by April 1 the year after a census year, therefore preventing the use of adjusted counts for purposes of redistricting (see National Research Council, 2004a:267).

Furthermore, the current approach to adjustment has a number of complications that continue to present a challenge to the production of high-quality estimated counts, including the quality of the data for movers (often missing or collected by proxy), matching errors, the treatment of missing data for nonmovers, the estimation of the number missed by both the census and the postenumeration survey, and the heterogeneity remaining after the use of poststratification of the match rate and the correct enumeration rate (resulting in correlation bias). This last objection will be reduced, but not eliminated, with the likely shift to the use of logistic regression instead of poststratification in 2010 (discussed below).

In addition, the use of adjustment is complicated since for some important applications one needs adjusted counts at low levels of demographic and geographic aggregation, and a sample survey, by design, is intended to make estimates at more aggregate levels. A decision whether to use adjusted counts for any purpose must therefore rest on an assessment of the relative accuracy of the adjusted counts compared with the census counts at the needed level of geographic or demographic aggregation. One key issue that depends on the application is whether to base this assessment on population shares or population magnitudes. The Census Bureau's decision not to adjust the redistricting data for the 2010 census, due for release by April 1, 2011, was based on the difficulty of making this assessment within the required time frame.

Census Process Improvement

Although it is important to assess census coverage, it would also be extremely helpful to use that assessment to improve the quality of subsequent censuses. Consequently, an important use of coverage measurement is to help to identify important sources of census coverage errors and possibly to suggest alternative processes to reduce the frequency of those errors in the future. Although drawing a link between census coverage errors and deficient census processes is a challenging task, the Census Bureau thinks that substantial progress can be made in this direction. Therefore, the 2010 coverage measurement program has the goal of identifying the sources of frequent coverage error in the census counts. This information can then be used to allocate resources toward developing alternative census designs and processes that will provide counts with higher quality in 2020. It is conceivable that use of such a feedback loop could also provide substantial savings in census costs, in addition to improvement in census quality because the tradeoff between the effect on accuracy and on census process costs might now be better understood. The panel fully supports this modification of the objectives of coverage measurement in 2010.

To see the value of this shift in the objective of coverage measurement, consider, for example, the findings from demographic analysis for the 2000 census, which showed that there was a substantial undercount of young children relative to older children. Specifically, Table 2-1 shows the net undercount rates—(DSE – C)/DSE, where DSE indicates the adjusted count, and C indicates the corresponding census count)—by demographic group in 2000 on the basis of the revised demographic analysis estimates (March 2001) (see National Research Council, 2004b:Chapter 6): One hypothesis is that the undercoverage for children aged 10 and under was at least in part due to the imputation of age for those left off the census

TABLE 2-1 Net Undercount in 2000

	Age Group	
Demographic Group	0–10	10–17
Black male	3.26	–1.88
Black female	3.60	–1.20
Nonblack male	2.18	–2.01
Nonblack female	2.59	–1.55

NOTE: The undercount is as measured by demographic analysis.
SOURCE: Data from National Research Council (2004b:Chapter 6).

form in households exceeding six members; this hypothesis is examined in Keller (2006). The 2000 census forms only collected characteristics data for up to six household members. For households that reported more than six members, characteristics data for the additional members either were collected by phone interview (for households that provided a telephone number) or were imputed on the basis of the characteristics of other household members and the responses for other households. The hypothesis is that these imputations systematically underrepresented young children since they were underrepresented in the pool of "donor" households.[2] Although demographic analysis can measure the net undercoverage of these groups, it cannot currently shed further light on the validity of this hypothesis. Data from a postenumeration survey might be useful in this regard, because characteristics data are collected for most of the respondents of the postenumeration survey, and those data would likely allow an assessment of the extent to which imputations in large households distorted the age distribution. Potential alternatives that could be considered for the 2010 census include changes to the collection of data for members of large households and improved imputation techniques.

The panel is optimistic that the use of coverage measurement can strongly support the improvement of census methods, but the operation of this feedback loop will not be straightforward. Coverage measurement results will sometimes provide strong indications of the likely source of some errors; for other errors, the source will often remain unclear. An example of the former is for people aged 18–21, who have a duplication rate that is extremely high: One might surmise that it is at least partly

[2]We note that even if it were determined that increasing this limit from six to seven would reduce the rate of omission of young children in large households, other considerations involving the rate of nonresponse and the quality of the collected information would have to be evaluated before making such a change.

due to the inclusion of college students in their parents' households as well as at college. For this situation, the process in need of modification is clear.

In contrast, a housing unit might be placed in the wrong location for many reasons, including erroneous coordinates for a MAF spot or a geocoding error using TIGER (the topologically integrated geographic encoding and reference database) and it may be difficult to specify the cause of the error. Many times such difficulties can be at least partly resolved through more detailed data analysis, assuming that additional information on the process history for the addresses in question is retained. Therefore, in moving toward the goal of process improvement, it is extremely important for the Census Bureau to save as much information on the procedural history for each housing unit and each individual within each housing unit and as much contextual information as possible, to develop useful statistical models linking enumeration errors to their possible causes. One possibility is to develop a comprehensive master trace sample database (see National Research Council, 2004a:Chapter 8) that is directly linked to the coverage measurement sample.

However, it is important to accept that there will always be limits to the attribution of errors to specific origins and therefore to the functioning of such a feedback loop. In particular, determining which alternative processes would best address a recognized deficiency would remain a challenge. For example, knowing that a geocoding error was the result of an error in TIGER does not necessarily tell one how to improve TIGER in a cost-effective way to eliminate that type of error. There are limitations to how well the feedback loop on census improvement can operate. In Chapter 5 we present some initial ideas on what data might be useful to save.

In sum, the Census Bureau has made an important shift in its focus for coverage measurement in 2010 from that of estimating net coverage error, potentially in support of census adjustment, to that of developing portions of a feedback loop for census improvement. However, as can be seen in detail in subsequent chapters, there remain vestiges of the previous goals in the design of and the outputs produced for the coverage measurement program in 2010. They include the sample design for the postenumeration survey in 2010, the current focus on the release of census tabulations as the main products of census coverage measurement rather than analytic uses of the collected data and the continued high priority of statistical models for net coverage error (the logistic regression modeling) in coverage measurement research.

Recommendation 1: The Census Bureau should more completely shift its focus in coverage measurement from that of collecting data

and developing statistical models with the objective of estimating net coverage error to that of collecting data and developing statistical models that support the improvement of census processes.

In order to ensure that the variety of issues identified in this report are addressed in support of improvements to the 2020 census design, it will be critical to have a team of high-quality researchers exclusively devoted to intercensal research on decennial census improvement and for this research program to be protected from year-to-year funding fluctuations and pressures. The activities of such a group would be focused on analyzing the data collected from the census, the census coverage measurement program, and the various predictors discussed below.

Recommendation 2: The Census Bureau should allocate sufficient resources, including funding and staff, to assemble and support an ongoing intercensal research program on decennial census improvement. Such a group should focus on using the data from the census and the census coverage measurement programs to identify deficient census processes and to propose better alternatives. The work of this group should be used to help design the census tests early in the next decade.

DESCRIPTION AND HISTORY

There are two primary methods that have been used for coverage measurement of the census: dual-systems estimation (DSE), supported by a postenumeration survey, and demographic analysis. To keep this document self-contained, we provide a brief description of these techniques and their history of use. For a more detailed description of DSE, see National Research Council (2004b:159–163, Chapter 6), and U.S. Census Bureau (2003). For a history of the U.S. census coverage measurement programs from 1950 through 1980, see National Research Council (1985: Chapter 4).

Dual-Systems Estimation

DSE uses both the data from the census and an additional enumeration to estimate the amount of net undercoverage in the census. The additional enumeration is typically a postenumeration survey (PES), which is a survey of the residents in a sample of census block clusters, who are referred to collectively as the P-sample. The PES, as its name indicates, is conducted after the main part of the census is completed on a housing unit by housing unit basis. In some cases, the census may still be ongoing for some households while the PES has initiated data collection for others.

The first step of a PES taking place, in recent times about a year prior to census day, is that the addresses in the PES blocks are independently listed, using no information from the MAF. (However, information on the differences between MAF counts and PES listing counts could be used in the sample design of the PES, since blocks in which the independent listing differed greatly from the count from the MAF might have been subject to a lot of recent construction or other recent dynamics that might have been a challenge to the census enumeration.)

After the census is completed, interviewers visit the P-sample housing units to establish which people were residents on census day. Additional information is also collected to support matching the P-sample results to the census and either to assign the persons to poststrata[3] (in the 1980, 1990, and 2000 coverage measurement programs) or to provide predictors for statistical models to estimate the total population size (see below). This additional information includes demographic and household characteristics and some area characteristics. For example, mailback rates and whether someone is an owner or renter, along with demographic characteristics, were used to define the poststrata in previous censuses. The purpose of the poststrata is to partition the P-sample into more homogeneous groups in terms of their coverage properties, so that when the coverage measurement is carried out separately by poststrata, the result is a reduction in a type of bias referred to as correlation bias (see below).

Once this initial data collection has been concluded, the P-sample enumerations are matched to the census enumerations to determine who in the P-sample was also counted in the census. People who cannot be matched to the census are reinterviewed to make any needed corrections due to discovered errors either in the data collection or the matching.

Estimation of Net Coverage Error[4]

We start from the implementation of DSE in the 1980 and 1990 censuses, which formally used the construct of a 2 × 2 contingency table as shown in Table 2-2:

- M is the estimate of the number of P-sample persons who match to a census enumeration (within the defined search area) in a poststratum, which will typically be a person in the E sample (the census enumerations in the P-sample block clusters,

[3]Poststrata are broad population groups defined by demographic or other characteristics.
[4]For a more comprehensive treatment of this subject, see Mulry and Spencer (1991).

TABLE 2-2 Diagram of the 2 × 2 Contingency Table Underlying Dual-Systems Estimation

	Counted in PES	Missed in PES	Total
Counted in Census	*M*		*C*
Missed in Census		Fourth Cell	
Total	*P*		*DSE*

NOTE: See text for discussion.

- P is the estimate of the number of all valid P-sample persons within a poststratum,
- C is the number of census enumerations in a poststratum, and
- DSE is the dual-systems estimate of the total number of residents within a poststratum, in other words the estimated true count.

The cell designated as the fourth cell is the only shaded cell in the above table that cannot either be directly measured or measured by subtraction using other directly estimated quantities, and it therefore has to be modeled and is assumption dependent, the assumption being that correlation bias, defined below, is acceptably small. It makes sense that this cell would be the problematic one since it counts those missed by both the census and the PES enumeration processes.

We assume initially for purposes of exposition that there are only omissions and therefore no sources of overcoverage. In addition, we assume that there are no missing data. Then the estimation of census undercoverage using DSE relies on two additional assumptions: (1) that the P-sample enumerations and the census enumerations are independent events, and (2) that all individuals within a poststratum have equal enumeration propensities. When these assumptions obtain to a reasonable extent, the following approximate equation will hold:

$$\frac{M}{P} \cong \frac{C}{DSE}. \qquad (1)$$

The above approximate equation can also be expressed as

$$DSE \cong \frac{C \cdot P}{M},$$

which suggests that one can estimate the total population size by taking the product of the number of census enumerations and the number of P-sample enumerations, divided by the number of matches.

The argument in support of the above approximate equation is as follows. Again ignoring some complications (discussed below), in a given poststratum, the first ratio in (1), (M/P), is an estimate of the percentage of census enumerations in the P-sample population, that is, an estimate of the census "capture" rate in the P-sample population (which, again, is assumed to be constant). The second ratio in (1), (C/DSE), is an estimate of the percentage of census enumerations in the full population, again an estimate of the census "capture" rate, but this time in the full population. If the P-sample selection and field data collection processes are independent of the census processes, and if the *operational independence* of the census and the PES also engenders *statistical independence*, then the fact of P-sample membership should provide no information as to whether a person was or was not enumerated in the census. Therefore, the population of P-sample enumerations should have the same underlying probability, conditional on poststratum membership, of being enumerated in the census as for the full population. Given that, these two ratios should be approximately equal.

As mentioned, the above argument assumes that each individual has the same chance of being enumerated in the census. However, the census and the P-sample enumeration probabilities are likely to be heterogeneous, i.e., dependent on various characteristics of the housing units and their residents. As a result, a bias (correlation bias) in the dual-systems estimate occurs, and its magnitude is a function of the degree to which the individual census enumeration propensities and the individual PES enumeration propensities are correlated. This correlation is small when either the census or the PES enumeration frequencies are relatively constant; therefore, if DSE is restricted to poststrata in which the people are relatively homogeneous with respect to their enumeration propensities, correlation bias will be relatively minor[5] (to see how enumeration heterogeneity can be related to dependence and therefore correlation bias, see Box 2-1). To address the complication of overcounts requires an additional data collection, which amounts to checking on the validity of the census enumerations. The E-sample, the census enumerations in the P-sample block clusters, is used to estimate the percentage of the census enumerations that are correct and not duplicated. This validation operation is carried out by visiting the E-sample people who fail to match to the P-sample to determine whether they were enumerated in error or in

[5]For a more detailed exposition of the errors in dual-systems estimation, see Alho and Spencer (2005:Chapter 10).

BOX 2-1
Example of the Relationship Between Enumeration Heterogeneity and Correlation Bias

To see how enumeration heterogeneity can be related to dependence and therefore correlation bias, consider three 2 × 2 tables: one for Area A, one for Area B, and the third the cell-by-cell sum of the first two.

Area A			
	Census In	Census Out	Total
PES In	25	15	40
PES Out	15	9	24
Total	40	24	64

Area B			
	Census In	Census Out	Total
PES In	24	3	27
PES Out	8	1	9
Total	32	4	36

Area A and Area B			
	Census In	Census Out	Total
PES In	49	18	67
PES Out	23	10	33
Total	72	28	100

The first two tables represent independence of the enumeration in the PES and the census. This is because, in each case, the fourth cell count is equal to the product of the probability of a census omission and the probability of a PES omission multiplied by the total number of individuals. For Area A, this is the equality: $9 = \left(\frac{15}{40}\right)\left(\frac{15}{40}\right)64$, and for Area B, this is the equality: $1 = \left(\frac{3}{27}\right)\left(\frac{8}{32}\right)36$. However, the equality does not hold for the combined area, since: $10 \neq 8.6 = \left(\frac{18}{67}\right)\left(\frac{23}{72}\right)100$.

So capture heterogeneity between Areas A and B has resulted in correlation bias for the combined area.

the wrong place and by matching the E-sample records to each other to check for duplicates. (The E-sample cases that match to the P-sample are validated as correct enumerations.)

This field operation will be augmented in 2010 with national computer matching to identify duplicate enumerations that are more remote geographically than those which jointly reside in the P-sample block cluster. This plan represents an important improvement when the goal is census improvement, since without it a PES cannot distinguish between a duplicate pair and a correct enumeration and another enumeration in the wrong place when one of the pair resides outside the P-sample blocks. In sum, the E-sample and the associated field work (and, soon, national matching) provide estimates of the percentage of people counted incorrectly in the census, the percentage of those counted more than once, and the percentage of those counted in the wrong place, which are used to adjust the above approach to accommodate overcoverage.

Finally, there are census enumerations that cannot be matched to the P-sample enumerations because they are not data defined. (For the specific definition of enumerations that are not data defined, see Chapter 4.) Since without being able to match, one cannot determine whether or not those individuals were or were not counted in the P-sample, one cannot determine whether the associated P-sample people were census omissions or correct census enumerations. These cases also need to be accommodated in DSE.

Using both the E-sample-based estimates of the rate of correct enumeration and the number of census enumerations that are not data defined, one can compute the following modification of the above dual-systems estimate.[6] C, the census count in the above formula is now replaced by

$$\left(C - II\right)\left(\frac{CE}{E}\right),$$

where II represents the number of non-data-defined enumerations,[7] CE represents the number of (survey weighted) E-sample persons correctly enumerated in the census, and E represents the (survey weighted) estimate of $C - II$. The number of people that are not data defined, II, is subtracted from the census count since their match or correct enumeration status cannot be determined. The implicit assumption is that their net coverage error is the same as that for the remaining census enumera-

[6]This discussion ignores the additional complication raised by the treatment of any late census additions, which are census enumerations that are too late to be included in the dual-systems computations.

[7]II can be interpreted to be the number of cases with insufficient information for matching, but as discussed in Chapter 4, many of those cases will be matched to the census in 2010.

tions (an assumption that may not be justified). Therefore, the number of matchable persons, $C - II$, is multiplied by CE/E, the percent of correct census enumerations, to estimate the number of data-defined census enumerations that are correct census enumerations. The resulting DSE formula is

$$DSE = (C - II)\left(\frac{CE}{E}\right)\left(\frac{P}{M}\right). \tag{2}$$

This derivation still ignores several additional nontrivial complications, including the treatment of other forms of missing data, the treatment of data from movers, and the precise area of search for matches of census enumerations outside the P-sample blocks. Additional complications arose in computing various revisions of the original estimates of undercoverage in the 2000 coverage measurement program (the Accuracy and Coverage Evaluation Program [A.C.E.]), due to the incorporation of information from various follow-up studies, though that is nonstandard and would not typically factor into the estimates of net coverage error. A description of the effect of these additional complications on the formulas used to produce the 2000 undercoverage estimates can be found in U.S. Census Bureau (2003).

In 1980 and 1990, the corrections generated by the E-sample were implemented by subtracting the number of cases insufficient for matching, duplicates, and erroneous enumerations directly from the census total, C, prior to use of formula (1). In the coverage measurement program used in 2000 and in the program planned for 2010, the estimation is no longer based on a 2×2 contingency table, but instead, the objective has become simply to provide estimates of the fractions CE/E and P/M for input into the formula (2). The estimates of population totals that result, assuming the use of poststrata, exist at the level of the individual poststrata. As noted above, these are groups of people that have similar characteristics related to enumeration propensity. For example, all have similar demographic characteristics, all are either renters or owners, have the same mail-back rate, and may share some high-level geography (e.g., the same census region).

Given the minimal representation of geography in poststrata, it is the case that DSE provides very little information about where people are missed. What is needed for many applications, rather than estimates of the net coverage error for poststrata, are estimates of the population for small political jurisdictions that are much smaller than the geographic level of the poststrata. In the 1990 and 2000 censuses, synthetic estimation was used for these estimates.

The basic idea of synthetic estimation is that the coverage correction factors,

$$\left(\frac{C-II}{C}\right)\left(\frac{CE}{E}\right)\left(\frac{P}{M}\right),$$

which are simply the dual-system estimates divided by the corresponding census counts at the level of the poststrata, are used to weight the census counts for any defined subpopulation of individuals within a given poststratum in order to produce an adjusted count for that specific subpopulation within that poststratum. Examples of subpopulations might be people of a specific geographic jurisdiction, or those from a specific demographic age-sex-race combination, or people belonging to both subpopulations. To arrive at an adjusted count for a geographic area of interest, one would produce the adjusted count for the geographic area of interest for all relevant poststrata, and then add those estimates together. An example of synthetic estimation is shown in Box 2-2.

Unresolved Aspects of Dual-Systems Estimation

The success of DSE depends to a great extent on the quality of the data collected, especially as to an individual's correct census residence. Both the decennial census data collection and the data collection for the PES inevitably involve misresponse and nonresponse.

Misresponse, which might stem from a misunderstanding of census residence rules, can cause errors in determining both match status and correct enumeration status. The matching errors that are made are quite likely more in the direction of false nonmatches than false matches.

Missing data complicate the application of DSE in the following ways. Although erroneous enumerations in the E-sample that have sufficient information for matching are typically identified as erroneous (though there are cases for which correct enumeration status has to be imputed), erroneous enumerations (and correct enumerations) that have insufficient information for matching are removed from the computation and are assumed to behave like the remaining E-sample enumerations in their poststratum through an implicit reweighting adjustment. The extent to which this assumption is valid can be examined using follow-up studies.

P-sample persons with sufficient information for matching that are missed in the census are typically correctly identified as nonmatches. However, P-sample cases with insufficient information for matching are accounted for by giving additional weight to those cases with sufficient information that are similar on available characteristics thought to be predictive of match status. The validity of these weighting (or possibly imputation) models can also be examined using follow-up studies.

As described above, the number of persons missed in both the P-sample and the census is estimated based on the assumption of independence of the two enumeration processes and of homogeneity of the enumeration propensities in poststrata (the absence of which engenders correlation bias). Although the use of poststrata is intended to partition the population into subgroups that have relatively homogeneous enumeration propensities, the poststrata are not completely homogeneous. Since no data are collected for this "fourth cell," it is unclear how well this subpopulation is estimated; merged administrative records might be used for this purpose. The general expectation is that this category of census omissions is underestimated. The difficulty in achieving homogeneity not only reduces the quality of the estimation of the fourth cell, but it also reduces the quality of any small-area estimates of net coverage error produced using synthetic estimation.

For duplicates, data-defined enumerations with name and date of birth in the E-sample in the P-sample block clusters are typically discovered, but duplicates outside the P-sample blocks were either undiscovered and erroneously treated as nonduplicate correct census enumerations or were categorized as erroneous enumerations. Assuming the national search for duplicates is implemented, these cases in 2010 will now have a chance of being identified as duplicates.

A limited search for matches also raises other difficulties. In previous censuses, when the corresponding census enumeration was located outside the P-sample block due to geocoding errors or other reasons, P-sample people that were not census omissions were designated as such. The frequency of this situation also should be reduced in 2010 with the implementation of the national search for matches. In the 1980, 1990, and 2000 censuses, the local restriction of the search for matches was designed to overstate the number of census omissions and the number of erroneous enumerations by the same amount. With a focus on estimation of net coverage error, the intention was that these errors would balance out, resulting in a zero net effect.[8] However, various deficiencies in the operation of the field processes may have upset this balance, leading to "balancing error." With the new objective of estimating the frequency of the various components of census coverage error, relying on a restricted search area for matches would result in a substantial increase in the estimated rates of omission and erroneous enumeration, much of which would be due instead to counting someone in the wrong location. Therefore, the current

[8]There is, however, an increase in the variance of the estimate of net coverage error due to the need for these random amounts to balance out for various domains.

BOX 2-2
Example of Synthetic Estimation

We present here a numerical example of synthetic estimation. Consider first the following fictional census count data:

Area	Demo Group I	Demo Group II	Demo Group III	Total
Area I	200	100	100	400
Area II	300	50	50	400
Area III	100	150	50	300
Total	600	300	200	1,100
Adjusted Total	800	300	300	1,400
Coverage Correction Factor	1.33	1.00	1.50	

The adjusted totals are provided by the coverage measurement program at the level of poststrata, which in this fictional example is simply represented as demographic groups. It is realistic to represent the adjusted totals as having no geographic structure relevant to this table because the sample size of postenumeration surveys are typically only large enough to provide reliable estimates at high levels of geographic aggregation. The problem that needs to be addressed is the 300 (1400 – 1100) "additional" enumerations that have been estimated to reside in the three areas and have to be allocated to the three areas to satisfy various uses of census counts. If one had an adjustment program that supported estimates at the level of the cells in the above table, one could take the adjusted count for each cell, and add across demographic groups to arrive at an estimated population count for each cell. This operation would be equivalent to estimating the coverage correction factor for each cell and multiplying that by the census count to estimate the count for each cell. However, a coverage correction factor for each cell does not exist.

plans to search nationwide for PES matches will be extremely helpful in arriving at a better understanding of the types and frequencies of components of census coverage error.

Lastly, errors in geography, or demographics, or other characteristics, can also result in the placement of individuals in the wrong poststratum, which can also bias the estimation of net coverage error.

DEMOGRAPHIC ANALYSIS

The Census Bureau has made substantial use of demographic analysis for several censuses, going back to 1940 (see, for example,

Underlying synthetic estimation is the assumption that the coverage correction factors are much more variable by demographic group than they are by area, so that using the coverage correction factors at the level of demographic group across areas, rather than disaggregated by area, may be of acceptable quality. This assumption is extremely difficult to verify empirically, though information from administrative records may provide some support in the future. In this fictional case, the assumption is that the factor 1.33 would apply relatively well to each of the three areas, and similarly for the factors 1.00 and 1.50.

Applying these coverage correction factors to the counts above results in the following synthetic estimates (adjusted cell counts) and the resulting column of estimates for each area:

Area	Demo Group I	Demo Group II	Demo Group III	Total
Area I	267	100	150	517
Area II	400	50	75	525
Area III	133	150	75	358
Total				1,400

The result is that 117 people are added to area I, 125 people are added to area II, and 58 people are added to area III. Tukey (1983) showed that synthetic estimates provide counts that are preferred to the census counts, regardless of where the additional people actually reside, for many typical loss functions when there is a single demographic group. The National Research Council (1985) showed that this optimality property does not extend to the situation in which one aggregates the estimates over demographic groups.

Price, 1947; Coale, 1955; Coale and Zelnik, 1963; Coale and Rives, 1972; Shryock and Siegel, 1973; Siegel, 1974). We present a short overview here; for a more detailed treatment relevant to the 2000 application, see Robinson (2001).

Basic Approach

Demographic analysis makes use of the following "balancing equation" to estimate the population in a demographic group: $P^{NEW} = P^{OLD} + B - D + I - E$, where by demographic group, we mean in particular groups defined by age, sex, and black or nonblack:

P^{NEW} is the current population for the demographic group;

P^{OLD} is the previous population for the demographic group, or can be zero when a population is started from births;

B is the number of births into the group occurring either from an initial date or from the date of a previous census;

D is the number of deaths to the group occurring either from an initial date or from the date of a previous census;

I is the amount of immigration to the group occurring from an initial date or from the date of a previous census; and

E is the amount of emigration from the group occurring from an initial date or from the date of a previous census.

As suggested in the above definitions, this equation can either be applied from a date prior to the birth of any of the current residents in that demographic group, assessing how many members of that group have been born here, remained here, and have not died, and how many members of that group have moved here, remained here, and not died; or it can be applied to assess the changes in the size of the demographic group that have occurred since a previous (often the most recent) census.

In addition to the use of data on births, deaths, and immigration, given their high quality, Medicare enrollment data are now used to estimate the population aged 65 and older without resorting to the above accounting equation.[9]

Unresolved Aspects of Demographic Analysis

The logic of demographic analysis requires that the population estimates constructed from the basic demographic accounting relationship be comparable with the populations measured by the census. As a result of demographic changes in the U.S. population over the past generation, many of the assumptions made by demographic analysis have become more problematic than they were for the censuses of 1940 through 1980. Specifically, immigration and emigration have become much more important sources of population growth. Also, as a result of immigration, intermarriage, and larger societal trends, the current definition and measurement of racial and ethnic groups have become less consistent

[9]The above equation is used in "reverse time" to backdate the Medicare-based population estimates to earlier censuses. For example, the population aged 55 and over in 1990 can be estimated by "reviving" the Medicare-based estimate for people aged 65 and over in 2000 adding deaths occurring in 1990–2000, subtracting 1990–2000 immigration, and adding estimated 1990–2000 emigration.

with historical definitions in the data used to construct the demographic estimates.

Historical data on the numbers of births and deaths are of relatively high quality; data on international migration are more problematic. Estimates of the number of emigrants are subject to considerable variability. In addition, undocumented immigration has become as important numerically as legal immigration, but the available measures are not very exact. Demographic analysis has generally been restricted to national estimates of age, sex, and race groups, since the available measures of subnational migration are not sufficiently reliable to support production of estimates at the state or lower levels of geographic aggregation. Furthermore, because ethnicity has been captured in vital records on a national scale only since the 1980s, demographic analysis has not been used to estimate net undercoverage for Hispanics. Finally, with the introduction of multiple-race responses in the 2000 census, it has become necessary to map the census race categories into historical single-race categories (or vice versa), with the attendant introduction of additional bias into the demographic analysis estimates. The introduction of multiple-race responses is part of the growing complexity of racial classification, which is likely to increase discrepancies between birth certificate reporting and self-reporting of race by adults.

Having pointed out some of the deficiencies of demographic analysis, it is important to emphasize its continuing value in coverage measurement. Demographic analysis places the census results within the well-defined, consistent, and essentially tautological framework of demographic change. The realities of the balancing equation shown above place severe limits on certain results from other coverage measurement programs. Thus, for example, with the passage of 1 year, all living people get exactly 1 year older; or, for every baby boy born, there will be approximately one baby girl born. If the results from other coverage measurement studies give results outside the bounds implied by such demographic realities, the results need to be explained. Some explanations may be demographic—for example, higher levels of immigration or emigration than included in the demographic estimate. But they may also point to statistical or measurement issues—for example, the persistent correlation bias that affects DSE measures of adult black men.

The current demographic analysis program at the Census Bureau also links the measures from the current census with censuses back to 1940. While DSE stands alone as a measure of the coverage of a particular census, the demographic analysis program grounds its current results in the historical data series, so that it is possible to assess one census relative to others.[10] This linkage can place limits not only on the demographic analysis measures, but also on the plausibility of results from other studies.

Due to the relatively deterministic nature of estimates of net coverage error from demographic analysis, estimates of its error or uncertainty are difficult to justify. However, there have been a few attempts to provide error estimates, in particular Robinson et al. (1993).

It is important to recognize that some results from demographic analysis are of higher quality than the overall results, so that they may be incorporated into a comprehensive coverage measurement program. While immigration has become an increasingly important component of population change that is not well measured, it has very little effect on the youngest age groups. Thus, it would be difficult to support estimates from DSE for children that were not consistent with demographic analysis results. Also, many population ratios, including sex ratios, are much less sensitive to assumptions about such problematic components as undocumented immigration than the measured population size. As a result, it may be possible to incorporate some byproducts of demographic analysis results into overall measures of coverage.

In 2000, demographic analysis proved to be very useful in the coverage measurement program, notwithstanding the noted deficiencies. Demographic analysis provided an early indication that the initial estimates of the total U.S. population from A.C.E. may have been too high. Also, demographic analysis has now been used to provide input to reduce the effects of correlation bias in DSE. Such corrections were used for adult black men in the 2000 revised estimates of net undercoverage and the current plans are to continue this use in the 2010 coverage measurement program (for more discussion, see Chapter 4).

1950–1990 CENSUSES

1950 Census[11]

The 1950 census used both a postenumeration survey and demographic analysis for coverage measurement. The strategy that motivated the postenumeration survey coverage evaluation methodology in the 1950 census was that coverage errors were largely due to failures to correctly implement

[10]Again, the realities of the balancing equation provide this linkage. Thus, if current research suggests that the demographic analysis coverage measures for a specific age group need to be adjusted (because, for example, the Medicare data show more or fewer enrollees than expected, or the births for a historical time period appear to be too high, or immigration during a decade had to have been higher or lower), the adjustment affects the size of the age group not only in the current census but also in past ones.

[11]Much of the material in this section is taken from National Research Council (1985).

census definitions and procedures and to imperfections in the materials and procedures used. Therefore, the idea was to take a sample of areas and in those areas "count again, but better." The 1950 PES used an area sample to identify whole household census omissions and a list sample from the census to estimate both within-household and whole household erroneous inclusions and to estimate within-household omissions. The list sample and the area sample overlapped to a great extent. Interviewers for the area sample canvassed each area, noted housing units not included in the list sample as possibly omitted from the census, and interviewed the occupants to collect census data. Interviewers for the list sample visited each household and determined any erroneous inclusions and any potential omissions, and in the latter case obtained census data from those individuals. All the interviewer records from both the area and the list samples were then matched to the census to directly determine the rates of erroneous inclusion and erroneous omission. To ensure high quality, interviewers were very carefully selected and trained. In addition, more detailed questions were asked, in comparison with the census, and interviewers were instructed to obtain responses from each adult rather than allow for proxy responses. The results were an estimated net undercount of 1.4 percent, with 2.2 percent omissions and 0.9 erroneous inclusions.

The demographic analysis in 1950 is described in Coale (1955). For people under 15 years of age, Coale used birth registration data to estimate their population size. For those aged 15–64, Coale relied on comparisons with the results of preceding censuses, subtracting estimates of the degree of mortality and adding net immigration for the various age groups in the intercensal periods. For those over 65, Coale used the 1950 postenumeration survey results to estimate net undercoverage. The results were an estimated national net undercount of 3.5 percent,[12] 2.5 times the PES estimate. The apparent ineffectiveness of the PES was attributed to the tendency for the PES to miss the same types of people who are missed in the census, i.e., to reflect correlation bias.

1960 Census

The 1960 census again used a PES that was composed of an area sample and a list sample (as in 1950). The area sample contained 2,500 segments comprising 25,000 housing units from the 1959 Survey of Components of Change and Residential Finance. Enumerators were instructed to canvass their areas and reconcile their lists with that of the survey and to interview the occupants of any omitted housing units. The list sample, independent of the area sample, comprised a national sample of 15,000

[12] Robinson and West (2000) give an estimate of 4.1 percent.

housing units drawn from the census enumerators' listing books. Interviewers were given the list of housing units, but not the names of the residents, and were instructed to re-enumerate the housing units. The results from both the area sample and the list sample were matched to the census. The resulting estimate of national net undercount was 1.9 percent.

Demographic analysis was also used to evaluate the coverage of the 1960 census. As described in Coale and Zelnick (1963), the method used was essentially an extension of the method used in 1950, though much more elaborate. Siegel (1974) utilized Medicare data to improve Coale and Zelnick's estimated number of elderly. The resulting estimate of national net undercoverage was 2.7 percent,[13] 1.4 times that of the PES.

In addition, the Census Bureau also carried out a reverse record check study to estimate net national undercount in 1960, which was very close to the methods used for many decades by Statistics Canada to assess net undercoverage. See Bureau of the Census (1964) and Gosselin and Theroux (1977) for details of this alternative approach to measuring net undercoverage.

1970 Census

As a result of the problems experienced with the PES in 1950 and 1960, the Census Bureau did not use a PES to evaluate the coverage of the 1970 census, placing primary reliance on demographic analysis. However, there was a CPS-census match study, whose estimates were adjusted for additions to the census count resulting from imputations based on the National Vacancy Check, and the Post-Enumeration Post Office Check. (The National Vacancy Check was in some sense the only use of sampling to adjust census counts, by selecting a sample to provide an improved estimate of how many housing units were vacant.) The Census Bureau also carried out two record check studies of specific population groups. The Medicare Record Check involved use of a sample of 8,000 persons age 65 and over that was selected from Medicare records and matched to the 1970 census. The D.C. Driver's License Study involved a match of driver's license records with census records for roughly 1,000 males aged 20–29 living in selected tracts in the District of Columbia.

The data used for demographic analysis estimates included birth and death statistics, immigration data, Medicare enrollments, life tables, and data from previous censuses. Siegel (1974) published a range of estimates, with a preferred estimate of national net undercoverage of 2.5 percent.[14] From 1950 to 1960 to 1970, there was increasingly less reliance on data

[13]Robinson and West (2000) give the estimate as 3.1 percent.

[14]Robinson and West (2000) give the estimate as 2.7 percent.

from previous censuses and more reliance on birth and death data, and, much less, data on immigration and emigration.

1980 Census

The coverage measurement program in 1980, which used DSE based on a postenumeration survey, was the Post-Enumeration Program (PEP). It was planned with the understanding that it might be used to adjust the 1980 census for net undercoverage. Therefore, the intent was to produce reliable net undercoverage estimates possibly down to the level of states and large cities. In contrast to the strategy used in the 1950 and 1960 PESs, the approach used in 1980 was "do it again, independently." The recount was not assumed to be superior to the census count, just independent. If this assumption obtained, DSE, described above, could then be applied to estimate net census undercoverage (as argued by Sekar and Deming, 1949, however, in the context of birth registration). The P-sample in 1980 comprised the April and August CPS samples (roughly 70,000 households each and about 185,000 individuals).

The two months of the CPS were used to provide estimates of sufficient reliability to support estimates of undercoverage for states and major cities. For the purpose of coverage measurement, the CPS interview was augmented with a sketch map to help locate the residence, and, for the August P-sample, a list of all recent residences of each person was requested. The census questionnaires were then searched (for only the relevant enumeration district) for a person with closely matching information on name, address, sex, age, and race. If no matching census record could be found, or if the information collected was insufficient, a follow-up interview was attempted.

The E-sample was designed to estimate the rate of erroneous enumerations, duplicates, and enumerations in the wrong place. The E-sample was composed of 100,000 census questionnaires, though only 50,000 were searched for duplicates. A substantial complication of the PEP was that the percentage of cases either not responding to the initial interview or whose matching status was unresolved was more than 8 percent nationally and was even larger for some demographic groups. Various treatments of the nonresponse and unresolved match cases and decisions about which CPS month to use (they were not used jointly to produce estimates) resulted in 12 different sets of undercoverage estimates, which varied considerably. For example, estimates of net national undercoverage ranged from 0.8 to 1.4 percent. Given this uncertainty, the Census Bureau decided against adjusting the 1980 census for differential net undercoverage.

Demographic analysis was also used to estimate the net undercoverage of the 1980 census. Passel et al. (1982) describes the methods used, which were similar to those used in the 1970 census. However, the rise in the population of undocumented aliens reduced the quality of the estimates from demographic analysis in comparison with those in 1970. Initial demographic analysis-based estimates of the net national undercount were very close to 0 percent. However, there is evidence that the initial demographic analysis estimates failed to account for a substantial increase in the undocumented immigrant population. Subsequent analysis suggests that the undercoverage of the census was as much as 1.5 percent given reasonable assumptions about the size of the uncounted undocumented population.[15]

1990 Census

Initial plans for the 1990 coverage measurement program were for a PES of 300,000 housing units, which the Census Bureau argued was needed to support net undercoverage estimates (and potentially adjustment) at the level of geographic aggregation consistent with such uses as reapportionment and redistricting. However, this design was rejected by the Secretary of Commerce and was replaced by a PES of 150,000 households, which was only to be used for purposes of coverage measurement and not for census adjustment. That decision precipitated a lawsuit that maintained the size of the PES at 150,000 households (ultimately, 165,000) but reopened the possibility of using the PES to adjust the 1990 census: the decision on adjustment, to be made by the Secretary of Commerce, would benefit from the deliberations of the members of a Special Secretarial Advisory Panel.

The 1990 PES was the first postenumeration survey with a survey instrument specifically designed for coverage measurement. The final design included more than 5,000 block clusters that were independently listed. (The design included people living in many types of group quarters residences.) The design also included a considerable amount of oversampling of blocks that contained a large fraction of historically hard-to-enumerate people. The P-sample residents were interviewed and matched to the E-sample, the census enumerations in the PES blocks, first using computer matching software, with difficult cases then subject to clerical review. Unmatched cases were followed up in the field. Given the development of a specific survey instrument, and due to better efforts to collect data, the percentage of nonrespondents and cases with unresolved match status was substantially lower than in 1980. Dual-systems estimates were constructed by separately computing net undercoverage estimates for 1,392 poststrata. Due to their high variance, these estimates were

[15]Robinson and West (2000) give the estimate of 1.2 percent.

smoothed using empirical Bayes regression methods. Finally, the estimates were carried down to the level of census blocks using synthetic estimation.

The problem of estimating both the number of undocumented aliens resident in the United States on census day and the percentage of those that were enumerated in the census posed more of a challenge to demographic analysis in 1990 than in 1980, given the larger size of the undocumented population in 1990. A "residual" process was developed to address this problem. The basic idea is as follows. An estimate of the number of legal foreign-born residents was developed using reported data on legal immigration. This figure was subtracted, at the national level, from the estimated number of foreign-born residents, either from the census long-form sample or from the CPS (and now, the American Community Survey), to arrive at an initial estimate of the number of illegal immigrants. This estimate was then inflated to account for undercoverage of this population (for details, see Robinson, 2001). Clearly, this required a few assumptions that were unlikely to hold, even approximately. However, there was, and still is, no preferred alternative methodology.

The Census Bureau released preliminary PES results in April 1991, estimating a national net undercount of 2.1 percent, with a difference of 3.1 percent between the rate of undercoverage of blacks and nonblacks. An internal Census Bureau group voted seven to two in favor of adjusting the 1990 census to remedy differential undercoverage. The Special Secretarial Advisory Panel split equally on the decision on adjustment. Ultimately, the Secretary of Commerce decided not to adjust the 1990 census.

2000 CENSUS[16]

Background

The initial planning for the 2000 census coverage measurement program was based on two assumptions: (1) that coverage improvement programs were unlikely to greatly reduce black–nonblack differential net undercoverage, and therefore the 2000 census was likely to continue an historical pattern of substantial differential net undercoverage of minorities, and (2) the time necessary to compute PES-based adjusted counts and validate them would make it extremely difficult to deliver adjusted counts in time for apportionment or redistricting of the U.S. House of Representatives unless substantial changes were made to the design of the census. Consequently, the Census Bureau initially decided to use

[16]Much of the material in this section is taken from National Research Council (2004b).

for its coverage measurement program in 2000 a strategy referred to as integrated coverage measurement. The idea of integrated coverage measurement was to limit the more extreme efforts at coverage improvement that delayed the PES data collection and to rely on the PES to address some of the coverage problems in the primary enumeration. The resulting PES counts, which would be the official census counts in a so-called "one-number census," would be of higher quality than the unadjusted census counts. It was anticipated that the execution of the PES would greatly benefit from the experience gained in carrying out the 1980 and 1990 PESs, and from the earlier (and therefore higher quality) data collection made possible by the elimination of late-stage coverage improvement programs. The integrated coverage measurement design called for a 700,000 household survey and accompanying matching operation, which would provide reliable *direct* estimates for states. (Direct estimates refers to each state's estimate of net undercoverage, which are based on the sample collected from households in that state alone.)

The plan to use integrated coverage measurement was jettisoned after the Supreme Court decision in January 1999 (*Department of Commerce v. United States House of Representatives*, 525 U.S. 316), which prohibited the use of sampling methods to produce counts for purposes of apportionment. This decision required the Census Bureau to greatly modify the design of the 2000 census as well as the associated coverage measurement program. With respect to the census itself, the Census Bureau was not allowed to use sampling for nonresponse follow-up as planned, since that would result in sample-based census counts.

This change, in turn, affected PES plans for the 2000 census because sampling for nonresponse follow-up ruled out using a particular version of PES, referred to as PES-B. In a PES-B, one attempts to enumerate in-movers in the P-sample blocks, assuming that their size and characteristics are roughly equivalent to those of the out-mover population. To determine whether those individuals were enumerated in the census, one determines the address at which they resided on census day, and then checks the census records at those locations to see if there is a match. However, with a census using sampling for nonresponse follow-up, many of those locations would not be included in the sample follow-up, or they would not have a census enumeration status, greatly complicating the use of PES-B. For that reason, PES-C was created, which again uses the size of the in-mover population to estimate the size of the out-mover population, as in PES-B, but instead uses the match status of the out-mover population. Because of the difficulty locating and contacting the out-movers, match status is often based on proxy information of dubious quality. For that reason, it is generally believed that PES-C is inferior to PES-B. However, when the design for the 2000 Census was revised as a result of the

Supreme Court decision of 1999, the Census Bureau did not have time to move back to a PES-B strategy.

The Accuracy and Coverage Measurement Program

The modified coverage measurement program, referred to as Accuracy and Coverage Evaluation Program (A.C.E.), was scaled back to a 300,000-household PES. Given the early work already carried out in selecting the integrated coverage measurement sample, the reduced A.C.E. sample was selected by subsampling from that sample. The smaller sample size for the A.C.E. could be justified since there was no longer a need to produce direct state estimates to support apportionment and so estimates from A.C.E. could borrow information across state boundaries. In addition, it is important to note that adjusted counts were not needed now until April 1, 2001, in support of redistricting, which provided additional time for validating the A.C.E. estimates.

As noted above, the 2000 A.C.E. sample was twice as large as the PES in 1990, with a P-sample of 300,000 households in 11,000 block clusters. Given the larger sample, there was less need to oversample, and the resulting variances of net coverage estimates were very likely substantially reduced in comparison with those for 1990. A.C.E. also used computer-assisted telephone and personal interviewing to facilitate the collection of P-sample data. A.C.E.'s dual-systems estimates used 448 poststrata (which were later collapsed to 416). This number, and the larger sample size, reduced the need, relative to 1990, for empirical Bayes smoothing.

In light of the possibility of adjustment (before the Supreme Court decision), the idea was that if A.C.E. could be demonstrated to provide valid estimates of net coverage error for poststrata and if the estimated net error differed appreciably by poststrata, then adjusted population counts from A.C.E. should be used for redistricting and for other official purposes. In practice, however, the Census Bureau's evaluations of A.C.E. discovered several problems. One problem was that A.C.E. estimated an overall net *undercount* of 1.2 percent while the initial demographic analysis estimated that the census had a 0.7 percent net *overcount*. Later revision of the demographic analysis estimates resulted in the estimate of a 0.3 percent net undercount, but this was still inconsistent with the estimate from A.C.E.

Other problems with A.C.E. concerned balancing error, the uncertain effects of a substantial number of late additions to the census, the level of error from synthetic estimation, the relative lack of duplicates identified by A.C.E., and the validity of whole-household imputations. These problems collectively led to a recommendation by the Census Bureau, seconded by the Secretary of Commerce on March 6, 2001, not to use A.C.E. counts for redistricting.

Post-2000 Research

A.C.E. Revision I

After the 2000 census, the Census Bureau carried out research to investigate the sources and magnitudes of error in the 2000 census, A.C.E., and demographic analysis. As part of this effort, the Census Bureau carried out two studies to collect additional information relevant to specific concerns regarding A.C.E. The purpose of the work was to determine the extent of person duplication, and, on a sample basis, to identify the correct residence for E-sample enumerations and to determine the correct match status for P-sample cases. The studies were the Evaluation Follow-Up Study, which involved reinterviewing a subsample of 70,000 people in E-sample housing units in 20 percent of the A.C.E. block clusters to determine correct residences on Census Day (with additional clerical review of 17,500 people who were unresolved) and the Person Duplication Studies, which involved nationwide computer matching of E-sample records to census enumerations using name and date of birth. This nationwide search permitted the first determination of the extent of remote duplication in the census, that is, cases in which the duplicated individuals did not both reside in the PES block cluster. The Census Bureau also examined the implementation of the targeted extended search (searches outside of the relevant P-sample block cluster for a match for situations in which there was a likely error in identifying the correct block) for matches to P-sample cases; estimation of the match rate and the correct enumeration rate for people who moved during the data collection for A.C.E.; and the effects of census imputations. As a result of these very detailed investigations, the Census Bureau judged that A.C.E. counts substantially underestimated the rate of census erroneous enumerations and hence tended to overestimate the true population size.[17] The Census Bureau subsequently released revised A.C.E. estimated counts on October 17, 2001, which were referred to as A.C.E. Revision I counts. Given concern stemming from the finding that there were considerably more errors of duplication than were originally estimated by the A.C.E., the Census Bureau again recommended that

[17]Fenstermaker and Mule (2002) estimated that there were a total of 5.8 million duplicates in the 2000 census. The problem could have been even larger, but the Census Bureau mounted an ad hoc operation early in 2000 to identify duplicate MAF addresses and associated returns, which removed 3.6 million people from the 2000 census.

these adjusted counts *not* be used, this time for the allocation of federal funds or other official purposes.

A.C.E. Revision II

Between October 2001 and March 2003, the Census Bureau undertook a further review of all the data collected in the census and the A.C.E. In addition, based on a comparison of the sex ratios from demographic analysis to those from the A.C.E., the Census Bureau decided to revise (increase) the A.C.E. estimates for males so that the resulting sex ratios were consistent with the sex ratios for demographic analysis for blacks 18 years old and older and for all other males more than 30 years old. This revision was based on the Census Bureau's belief that the A.C.E. counts had been reduced by correlation bias. The adjustment that was implemented is implicitly based on the assumption that the correlation bias for the parallel female groups was close to zero. The estimates based on these revisions are referred to as A.C.E. Revision II. Evaluations of the quality of these final A.C.E. estimates resulted in the announcement, on March 12, 2003, by the Census Bureau that the A.C.E. Revision II counts would not be used as the base for producing intercensal population estimates.

The Panel to Review the 2000 Census (National Research Council, 2004b) generally agreed with the decisions made at each stage of this three-stage process, namely: not to use the A.C.E. counts—either the original, Revision I, or Revision II—for purposes of redistricting, fund allocation or other official purposes or for intercensal estimation. However, the panel was not in complete agreement with the supporting arguments of the Census Bureau.[18]

As a by-product of this intensive effort to understand whether adjusted counts were preferable to unadjusted counts for various purposes, the Census Bureau produced comprehensive documentation and evaluation of the A.C.E. processes. A considerable amount of material is available at the following locations. Evaluations supporting the March 2001 decision can be found at http://www.census.gov/dmd/www/EscapRep.html; evaluations supporting the October 2001 decision can be found at http://www.census.gov/dmd/www/EscapRep2.html; and evaluations supporting the March 2003 decision can be found out at http://www.census.gov/dmd/www/ace2.html. Collectively, these reports document the A.C.E. procedures in detail, examining what was learned about the quality of A.C.E. and A.C.E. Revisions I and II through the additional information collected.

---

[18]For the panel's arguments, a more detailed description of A.C.E. and the various evaluation studies, and the material on which this abbreviated history is based, see National Research Council (2004b:Chapters 5–6).

Limitations of A.C.E.

The A.C.E. in the 2000 census was planned from the outset as a method for adjusting census counts for *net* coverage error so it did not focus on estimating the number or frequency of the various components of census coverage error. As an important example, the limited geographic search for matches in the A.C.E. (for estimation of net coverage error) relied on the balancing of some erroneous enumerations with omissions, in which the erroneous enumerations were at times valid E-sample enumerations but in the wrong location. People can be counted in the wrong location as a result of a geocoding error (placing an address in the wrong census geography) or enumerating a person at a second home. Because such erroneous enumerations and omissions were expected to balance each other on average, they were expected to have little effect on the measurement of net coverage error. Therefore, the A.C.E. did not allocate the additional resources that would have been required to distinguish these situations from enumerations in the wrong place. Similarly, the A.C.E. did not always distinguish between an erroneous enumeration and counting a duplicate enumeration at the wrong location.

In addition, the A.C.E. effectively treated all cases with insufficient information for matching as imputations although it is clear that a substantial fraction of them are correct. The Census Bureau has done research that demonstrates that for a substantial subset of the cases match status can be reliably assessed. (See Chapter 4 for the discussion on missing data methods for details on this research.)

We now mention several other limitations of the A.C.E. in 2000 for measuring census component coverage error, along with any plans to address the limitation in the designs for the census and census coverage measurement in 2010. However, we must stress that our treatment here of the theoretical underpinnings of A.C.E. is incomplete. For a complete description of this, see Hogan (2003).

First, inadequate information was collected in the census on census day residence. In 2000, comprehensive information was not collected from a household in the census interview regarding other residences that residents of a household often used or on other people who occasionally stayed at the household in question. This limited the Census Bureau's ability to correctly assess residency status for many people. The Census Bureau intends to include more probes to assess residence status in the 2010 census questionnaire, and the coverage follow-up interview will also collect additional information on residence status for those housing units that are likely to have incorrectly represented the number of residents on the census form. In addition, more probes about residence will also be included on the 2010 census coverage measurement questionnaires.

In 2010, the duplicate search will be done nationwide, and not only for the PES population, to help determine census day residence. In conjunction with search, as part of the coverage follow-up interview, the Census Bureau plans on incorporating a real-time field verification of duplicate enumerations in 2010. (For details on issues in determining correct residence, see U.S. Census Bureau, 2003.)

Second, nonresponse in the E- and P-samples complicated matching of the P-sample to the E-sample and of the E-sample to the census (to identify duplicates). It also complicated estimation because it interfered with assigning a person to the correct poststratum (for details, see Mulry, 2002).

Third, the use of the methodology for individuals who moved between Census Day and the day of the postenumeration interview (known as PES-C; see above) resulted in a large percentage of proxy enumerations, which in turn resulted in matching errors. The Census Bureau is planning on returning in 2010 to the use of PES-B (similar to the 1990 methodology), which relies completely on information from in-movers.

Fourth, the A.C.E. Revision II estimates refined undercoverage estimates for black men over 18 and for "all other men" over 30 using sex ratios from demographic analysis (ratios of the number of women to the number of men for a demographic group) to correct for correlation bias (for details, see Bell, 2001; Shores, 2002). This method assumes that the net coverage error for women for the relevant demographic group is ignorably small. However, for nonblack Hispanics, refinement (other than that used for all nonblack males) using sex ratios would require a long historical series of Hispanic births and deaths and, more importantly, highly accurate data on the magnitude and sex composition of immigration (both legal and undocumented). Yet the historical birth and death data for Hispanics are available only since the 1980s, and the available measures of immigration are too imprecise for this application. Consequently, this use of demographic analysis to refine A.C.E. estimates was not directly applicable to nonblack Hispanic males in 2000.[19] In addition, there is a great deal of uncertainty about the degree to which various assumptions need to obtain to support the use of this methodology for either blacks or nonblack Hispanics. (The decision could also depend on the yardstick in question, i.e., whether counts or shares are the quantities of interest.)

Fifth, poststratification is used to reduce correlation bias since it partitions the U.S. population into more homogeneous groups regarding their enumeration propensities. The number of factors that could be included in

[19]For example, it is noteworthy that about 55 percent of working-age (18–64) Hispanics are foreign born in comparison with less than 5 percent of whites and slightly more than 5 percent of blacks.

the poststratification used in the A.C.E. was limited because the approach (essentially) fully cross-classified many of the defining factors, with the result that each additional factor greatly reduced the sample size per poststratum; for details of the 2000 poststrata, see U.S. Census Bureau (2003). The 2010 plan is to use logistic regression modeling to reflect the influence of many factors on coverage rates.

Sixth, small-area variations in census coverage error that are not corrected by application of the poststratum adjustment factors to produce estimates for subnational domains (referred to as synthetic estimation) were not reflected in the variance estimates of adjusted census counts. The Census Bureau is examining the use of random effects in their adjustment models to account for the residual variation in small-area coverage rates beyond that which is modeled through synthetic estimation.

Finally, the sample design made the 2000 A.C.E. less informative than it might have been in measuring components of census coverage error by not providing a greater sample targeted on people and housing units that are difficult to enumerate. As noted above, this approach was understandable given the focus of the A.C.E. on producing adjusted census counts. However, given the new priority of measuring the components of census coverage error, a number of design and data collection decisions in the general framework of PES data collection, especially sample design, remain open to modification.

The Role of Demographic Analysis

From 1950 through 1980, demographic analysis served as the primary source of estimates of national net coverage error in the decennial census. However, demographic analysis is fundamentally limited: It can only provide estimates of net coverage error for some national demographic groups. Demographic analysis cannot yet provide estimates for any geographic region below the national level, and it does not provide estimates of net undercoverage for the Hispanic population. Because of these limitations, DSE, based on a postenumeration survey, displaced demographic analysis as the primary coverage measurement instrument, starting with the 1990 census.

However, its limitations do not mean that demographic analysis is no longer useful. As was evident in 2000, demographic analysis still performs two useful functions. First, it provides estimates of population size that can be used in a variety of ways to assess the quality of dual-systems estimates, including comparison of dual-systems estimates to demographic analysis estimates that are less affected by the quality of the data on external migration and sex and through age ratios. Demographic analysis played this important role in the 2000 census. Second, sex ratios and other

TABLE 2-3 Net Undercoverage for U.S. Censuses, 1940–2000

Group	1940	1950	1960	1970	1980	1990[a]	2000[b]
U.S. Total							
Demographic analysis	5.4	4.1	3.1	2.7	1.2	1.8	0.1
PES	—	—	1.9	—	0.8–1.4	1.6	–.5
Black							
Demographic analysis	8.4	7.5	6.6	6.5	4.5	5.7	2.8
PES	—	—	—		5.2–6.7	4.6	1.8
Nonblack							
Demographic analysis	5.0	3.8	2.7	2.2	0.8	1.3	–0.3
PES	—	—	—	—	—	—	—

[a]Revised estimates.
[b]Revised demographic analysis and final A.C.E. estimates.
SOURCE: Data from Anderson (2000) and National Research Council (2004b).

information from demographic analysis can be useful in improving the estimates from DSE. For example, the modification of the count for adult black men, mentioned above, remains a possibility for 2010. In addition, looking to the future, ongoing research is attempting to address the two primary deficiencies of demographic analysis—the lack of subnational estimates and the lack of estimates for the Hispanic population.

To complete this history of coverage measurement, Table 2-3 shows the national estimates of net coverage error and the estimates disaggregated by black and nonblack for the decennial censuses from 1940 to 2000.

3

Plans for the 2010 Census

The 2010 census has an innovative design, resulting in a census that will differ from its predecessor to a very substantial degree. Though plans for the 2010 census remain tentative, it is useful for the panel's analysis to be able to compare the timetables for main activities in the 2000 and in the 2010 censuses. Box 3-1 contains a cross-walk of the timetables for the 2000 and 2010 censuses.

MAJOR DESIGN CHANGES

Four significant differences in design will considerably affect how the 2010 census coverage measurement (CCM) program needs to differ from the 2000 coverage measurement program: a short-form only census, an improved system for the Master Address File (MAF/TIGER) (the topologically integrated geographic encoding and reference database), coverage follow-up interviews, and removal of duplicate enumerations during the census.[1]

A Short-Form Only Census Since 2005 the Census Bureau has been fielding the American Community Survey (ACS), a continuous version of the decennial census long form. Therefore, under current plans there will be no long form in the 2010 census. This change will facilitate several

[1]The discussion in this chapter is based on the Census Bureau's plans for the 2010 census as of spring 2008.

BOX 3-1
Cross-Walk of Schedules

		2000 Census	2010 Census
LUCA[a]		LUCA 98 05/98–09/99 LUCA 99 01/99–10/99	Ship materials 11/06/07–3/18/08 Updates: 9/25/07–10/08/08 [Note: Some materials sent earlier than 11/06/07]
MAF Block Canvass		01/99–05/99	04/06/09–07/10/09
Questionnaire Mailout		03/13/00–03/15/00	03/15/10–03/17/10
NRFU[b] Begins/Ends		04/00–07/00	05/01/10–07/10/10
CEFU[c]	CFU[d]	05/00–07/00	04/26/10–08/13/10
CIFU[e]		07/00–08/00	
Coverage Measurement Personal Interviews		05/00–08/00	08/14/10–10/02/10

[a]LUCA: Local Update of Census Addresses
[b]NRFU: Nonresponse Follow-Up
[c]CEFU: Coverage Edit Follow-Up
[d]CFU: Coverage Follow-Up
[e]CIFU: Coverage Improvement Follow-Up
SOURCES: Census 2000 Operational Plan, December 2000, U.S. Department of Commerce, Economics and Statistics Administration, U.S. Census Bureau; 2010 Census Key Operational Milestone Schedule.

aspects of data collection in the census, including data capture, the work of follow-up enumerators, the management of foreign language forms and foreign language assistance, and data editing and imputation for nonresponse.

The ACS has been mentioned as a possible survey vehicle for coverage measurement. We agree that there may be some potential for use of the ACS to help assess the quality of dual-systems estimation (DSE), or to help more broadly in coverage evaluation. However, some problems would need to be overcome in applying the ACS in this way. First, the address files for the ACS and the Census are very closely related, so at present the ACS could not be used to estimate whole household omissions. In addition, the ACS questionnaire is not focused on coverage measurement, as is that for the CCM. Finally, the ACS has a different definition of residence than the census, which would cause some additional, albeit minor, complications.

Improved MAF/TIGER System In outline, the MAF begins with the final address list developed in concert with the taking of the previous census. This list is updated on a fairly continuous basis by additions and deletions to the U.S. Postal Service's Delivery Sequence File. For the 2000 census, local areas were provided the opportunity in 1998 and 1999 to make additions and deletions based on local information, which was referred to as the Local Update of Census Addresses (LUCA) Program. A block canvass was carried out to determine the accuracy of the address list a year prior to the census. In addition to these procedures, there were more than a dozen other ways in which an address can be added to the MAF. There were numerous questions about the completeness and the accuracy of the MAF listings for the 2000 census (see, e.g., National Research Council, 2004b:Finding 4.4), and efforts are now under way to improve both MAF and TIGER for 2010.

The MAF and TIGER databases have been redesigned into a single MAF/TIGER database: MAF provides a list of household addresses, and TIGER is used to associate each address on the MAF with a physical location. The MAF/TIGER Enhancement Program includes: (1) the realignment of every street and boundary in the TIGER database; (2) development of a new MAF/TIGER processing environment and the integration of the two previously separate resources into a common technical platform; (3) collection of global positioning system coordinates for structures on MAF; (4) expansion of geographic partnership programs with state, local, and tribal governments, other federal agencies, the U.S. Postal Service, and the private sector; (5) implementation of a program to use ACS enumerators to generate address updates, primarily in rural areas; and (6) the use of periodic evaluation activities to provide quality metrics to guide corrective actions (for details, see Hawley, 2004). One motivation for this initiative was the recognition by the Census Bureau that many census errors and inefficiencies in 2000 resulted from errors in the MAF and in the information on the physical location of addresses.

Coverage Follow-Up Interviews The Census Bureau is greatly expanding the percentage of housing units for which there will be a coverage follow-up interview in 2010 in comparison with the housing units in 2000 for which there was a coverage edit follow-up. The 2000 coverage edit follow-up was used to determine the correct count and characteristics for two situations: households with more than six residents (since the census form had space for information for only six persons) and households with count discrepancies (e.g., differences between the number of separate people listed on the questionnaire and the indicated total number of residents). The planned expansion in 2010 is motivated by the recognition that confusion about residence rules was a substantial source of census coverage error.

The expanded coverage follow-up interviews planned in 2010 will include four situations in addition to the two covered in 2000: (a) households with a possible duplicate enumeration identified by a computer match of the census returns to themselves; (b) households that respond positively to a coverage probe on the census questionnaire concerned with census omissions; (c) households that respond positively to a coverage probe on the census questionnaire concerned with census erroneous enumeration and duplication; and (d) households that have different counts than that of a census-developed population register based on merged administrative records, known as StARS.

Of these four situations, (a) is intended to identify both households containing duplicated individuals and fully duplicated households, (b) is intended to identify potential omissions in the census, (c) is intended primarily to identify duplicated individuals, and (d) is intended to identify all types of coverage error; see Box 3-2.

BOX 3-2
Situations Potentially Generating a Coverage Follow-Up Interview

Coverage follow-up interviews could result from any of the following six situations:

1. Count discrepancies in which the indicated total number of residents does not equal the number of individuals for whom information is provided on the census questionnaire.
2. Large households, where the number of residents is larger than six, which is the maximum number of individuals with space for characteristics information on the census questionnaire.
3. Positive result from the national duplicate search, i.e., where individuals in a household unit match the data for individuals in another household.
4. Positive response to the coverage probes for census omissions, namely: "Were there any additional people staying here *Census Day* that you did not include in Question 1?"
5. Positive response to the coverage probe for census overcounts, namely: "Does person P sometimes live or stay somewhere else?"
6. A count discrepancy between the census count for a housing unit and the count from a roster produced from merging administrative records.

Due to resource and time constraints, the Census Bureau will only follow-up those households that provide a telephone number. In addition, the Census Bureau may only be able to administer the coverage follow-up interview to a "most promising" subset of the qualifying households in 2010. In other words, the Census Bureau may have to set priorities by selecting a subset of the qualifying households that are more likely to provide information that would result in coverage improvements.

Removal of Duplicate Enumerations During the Census As noted above, the coverage follow-up interviews will be used to collect more information on suspected duplicate enumerations that are identified through use of a national computer search, with the objective of determining whether they are in fact duplicates and, if so, which of the addresses (if either) is the correct residence. If the correct residence is identified, the enumerations at the incorrect residence would be removed from the census.

This new census design has some benefits for the coverage measurement program in 2010. Focusing on the collection of short-form data will likely improve the quality of the information collected, thereby reducing the frequency of errors made in matching of the postenumeration survey (PES) to the census. Also, implementing a national search for, and field verification of, duplicate enumerations should reduce the number of duplicates in the census, which may in turn facilitate the estimation of components of coverage errors in the census and may also simplify the application of the net coverage error models used in DSE in 2010.

TREATMENT OF DUPLICATES

Census Duplications

There are many different causes of duplication in a census. As noted above, the census process may enumerate people that move either shortly before, on, or shortly after Census Day at both their previous and their current residences; the census may enumerate families with second homes at both residences; the census may enumerate college students both at their college residence and at their parents' homes; and the census may enumerate "snow birds" at both their primary residences and at their winter homes. These are all examples of confusion over where someone's correct census residence is.

Another cause of duplication in the census is representation of an address in more than one way on the MAF or having two forms returned for the same unit, which can happen in multiple ways. To address duplication in the census (in addition to attempts to measure its frequency in the coverage measurement program) efforts have been made to adjust various census processes to reduce the frequency of duplication. An example is the primary selection algorithm, used in both the 1990 and 2000 censuses and planned for use in 2010, which removes duplicate responses in the census from the same housing unit by identifying the unique people who were enumerated across all responses keyed to that housing unit. Also, the census questionnaire has been adjusted in attempts to reduce misunderstandings of census residence rules—in particular, through the addition of the two coverage probes—and various efforts have been made to reduce duplication in the MAF. Yet preventing census duplication before it occurs is still a nontrivial task, and it was a serious problem in 2000 (for details, see National Research Council, 2004b).

As noted above, the Census Bureau in 2010 will attempt to identify and delete duplicate persons and housing units during the census. Specifically, after the primary enumeration process and nonresponse follow-up are complete, a nationwide computer search of census enumerations for matching individuals will be carried out, using name, date of birth, gender, and phone number (when available). On the basis of the results of that search, the Census Bureau will identify likely duplicates in the 2010 census. Depending on the geographic proximity of the two residences in question and the duplicate status of the other residents of a household, this process may also be used to identify suspected duplicate housing units.

Once this list of potential individual and whole household duplicates is generated, the plan is to collect more information by telephone through coverage follow-up interviews at both residences. The interviews will attempt to ascertain which (if either) of the two enumerations is correct and which is a duplicate. (See Box 3-1 for additional details on circumstances that can generate a coverage follow-up interview.)

Because Title 13 (of the U.S. Code) privacy protections prohibit using information from one housing unit in querying another, extensive probes will be used for handling a wide variety of complex living situations that may be associated with the potential duplication in question (e.g., part-time residents, students away at college, movers, children in joint custody, and elderly people in nursing homes). In particular, the interviewers cannot be told which people in a house are likely duplicates to help guide the interviews. To reduce costs, as noted above, the coverage follow-up interviews will only be carried out by telephone, and so will not be carried out for households that did not provide their telephone number on their census questionnaires. This approach will prevent a modest, but not insubstantial, percentage of the existing duplicates from being identified and therefore being removed from the census.

Though the specific algorithm and the accompanying threshold for designating matching individuals have not been chosen, the Census

Bureau intends to set a strict threshold before the records for two individuals will be identified as a possible match and therefore trigger a coverage follow-up interview. In addition to the strict threshold for designating potential matches, one of the enumerations of a potential duplicate pair will not be deleted from the census unless the evidence collected from the coverage follow-up interview is clear that the individuals are duplicates *and* which of the residences is correct given the census residence rules.

The panel is unclear precisely how the information collected in the coverage follow-up interview will be used to discriminate between a duplicate enumeration and a nonduplicate enumeration and to determine the correct census residence. Moreover, the process will provide an asymmetric treatment of coverage errors in that the error that results in the removal of a valid enumeration will be judged as being more serious than the error that results in the retention of a census duplicate. The panel acknowledges that this asymmetry can be partly supported given the nature of decennial census counts: that is, because the political environment in which the census operates reacts differently to these two types of error. However, both errors need to be measured and the trade-off evaluated to determine if it is reasonable or needs to be reconsidered.

Joint Custody Scenario. As an example of what might happen in 2010, consider the following situation involving a child in joint custody, when both parents consider the child's primary residence to be her or his home. In this situation, the coverage follow-up interviews might well collect information that supports the same two residences as the census reported. The Census Bureau will strongly suspect that it is a duplicate pair but will be unable to delete either enumeration given the lack of a way to identify the correct residence.

Given the political sensitivity of the deletion of a census enumeration, a coverage follow-up interview is required for deletion of one of a duplicate pair, even if the duplicate status is essentially unambiguous given the above matching characteristics, and even if the residence rules are clear as to which residence is correct.

In the case of potential whole housing unit duplicates, field inspection will be used to determine if two housing units are duplicates. Potential duplicate housing units (or households) may result from: (1) duplicate addresses for the same housing unit, (2) delivery mix-ups in apartments, (3) movers, and (4) person duplication of all members of a housing unit. When field follow-up is used to verify duplicate housing status, there will be no associated telephone coverage follow-up interviews for the individual residents. If duplicate addresses are discovered for the same physical unit, one of the two enumerations will be deleted. In the case of a delivery mix-up, the duplicate is retained as a field imputation.

The Census Bureau's proposed use of the coverage follow-up interviews and the field validation of potentially duplicated whole housing units raises at least two major questions:

1. Given that the coverage follow-up interviews will be by telephone only, what are the anticipated effects on duplicate resolution and other enumeration problems that are addressed by the use of the interview?
2. What are the rules for deleting whole housing units that are identified as potential duplicates?

The coverage follow-up interviews and the national search for duplicates could provide substantial benefits over previous censuses in identifying and removing many census duplicates during the field enumeration and in reducing the occurrence of other census coverage errors. However, there are many potential complications in implementation that might limit the benefits from the introduction of these processes in 2010. Since these activities are inherently national in scope, they could not be comprehensively tested in the relatively limited environment of a census test or the 2008 census dress rehearsal. In particular, such environments are unlikely to provide very good estimates of the extent to which these new activities will stress the census infrastructure (e.g., because of the number of coverage follow-up interviews that will be required). However, they could provide information as to how to set the threshold for determining when potential duplicates have characteristics that are close enough to warrant a coverage follow-up interview.

The panel sees three concerns for the planned coverage follow-up interviews. First, will there be sufficient resources to support the interviews for all the situations that have been identified as potentially requiring such follow-up. Second, although the questions on the coverage follow-up interviews are more detailed than those on the census, will the similarity of the questions result in the relatively infrequent collection of information that would support changes in census enumeration status. Note that, in some sense, a respondent in the follow-up interview needs to admit that the previously provided information on the census form was incorrect. Third, given that follow-up interviews will only be done for households that provide telephone numbers, what will be the effects of not following up households that did not provide telephone numbers. In addition, as noted above, the panel is concerned that because the threshold for matching will be set relatively high, some duplicates will not appear to have characteristics that match and therefore will not trigger a coverage follow-up interview. This is particularly noteworthy because there may be demographic groups or geographic areas with a concentration of duplicates.

Careful evaluation of both the coverage follow-up interviews and the national search for duplicates is extremely important so that the functioning of these processes in 2010 is fully understood and to carefully guide any needed improvements for both of these processes prior to their use in 2020 (if, as we would anticipate, they are included in the 2020 census design). To support a careful evaluation, there is a need, at least for a sample of cases, to retain information as to precisely what happened to the cases that were selected for coverage follow-up interviews and as a result of the interviews. Therefore, the Census Bureau should save the responses, at least for a sample of enumerations, to the coverage follow-up interview questions and the final decisions made regarding the assessment of enumeration status.

In addition to retaining information on the functioning of the follow-up process, it will also be important to know the extent to which the follow-up process moved census counts closer to the truth. The CCM provides a unique resource to assist in determining the situations for which the coverage follow-up process worked well and those for which it worked poorly. So, for enumerations in the E-sample (census enumerations that are in the P-sample, the postenumeration survey clusters), it would be very useful to retain a comprehensive log of their status prior to and after the coverage follow-up interviews. With this information, the CCM can provide a formal way of measuring the probabilities of proper and improper duplicate removal and proper and improper duplicate retention, and it can therefore provide an assessment of the decision process that was used to determine the cases that were deleted as duplicates and the cases that were retained.

In addition to using the CCM for this purpose, it may also be valuable for the Census Bureau to return to the field to examine a subsample of cases selected for coverage follow-up interviews to see whether the interviews actually provided new information with any appreciable frequency and whether that new information led to correct decisions. Such a study should be designed to include a large fraction of census duplicates.

Regarding the national search for duplicates, it would be useful to learn more about those cases that were near but still below the threshold and therefore were not selected for coverage follow-up to determine whether other thresholds would have provided better results. One could sample from the cases near but below the threshold and follow them up to assess whether any were duplicates and whether a field interview likely would have determined that. Such data collection would inform a cost-benefit analysis of the tradeoff of identifying more true duplicate enumerations against the cost of the additional field work (and the erroneous identification of more false duplicates).

This discussion has been focused on the use of the coverage follow-up interviews to determine duplicate status. However, the interviews will also be generated by a coverage probe for census omissions (i.e., "Were there any additional people staying here *Census Day* that you did not include in Question 1?"). This probe is likely to have the same problem as that encountered in the search for duplicates, namely, that it will often result in the same information as the census. Therefore, it also will be important to evaluate the resolution of households whose selection for follow-up interviews are generated by the coverage probe for census omissions. These evaluations will help measure the degree to which the panel's concerns (noted above) are a problem. Looking ahead, to further improve on the use of this new process, the panel also believes it is important for the Census Bureau to undertake research during the postcensal period in the following areas:

- the potential for StARS to help target the cases that are included in the set of coverage follow-up interviews;
- the potential for StARS to help resolve potential duplicate enumerations;
- the potential use of StARS to augment the CCM personal interviews for resolving duplicate status;
- how to optimally set the "bar" for inclusion in the coverage follow interviews;
- how best to discriminate between person and whole household duplication; and
- in general, how to evaluate census unduplication procedures.

With respect to the first three issues above, see further discussion below. Given the late date, it may be difficult to comprehensively evaluate these three suggestions, but it should be feasible to make some progress on each.

CCM Duplications

Duplications will occur not only in the census but also in the CCM survey data collection. There are some differences in the treatment of a possible duplicate enumeration in the census and in the CCM. Some of these differences in approach are due to the dramatically different sizes of the two activities: coverage follow-up interviews could be used for between 10 and 30 million households; in contrast, the CCM will cover about 300,000 households. We note several consequences of these differences.

First, since one of the components of census coverage error is omissions, and since, as pointed out below, the estimation of net coverage error is needed to estimate the number of census omissions, the Census Bureau should and will use an unbiased approach to its assessment of duplicate status in the CCM, in the sense of avoiding any differential bias in assessing the number of census omissions in comparison with the number of census overcounts. Second, because the CCM does not have to make a final determination of enumeration status, as the census does, the CCM can assign probabilities of being a duplicate to cases with unresolved duplicate status or similarly to cases with unresolved correct enumeration status in the E-sample or unresolved residency status in the P-sample.

The in-person follow-up interviews for initially nonmatching CCM cases should often prove useful in reducing the number of duplicates in the CCM P-sample. However, it would also be useful to use information from the coverage follow-up interviews to reduce duplicates in the P-sample. However, due to concerns about statistical independence between the census and the postenumeration survey data collections, the Census Bureau does not currently plan to use information from the census coverage follow-up interviews to help ascertain duplicate status in the CCM. The panel does not agree with this decision: the panel does not understand why census information should not be used to assist in such determinations, since the goal is the proper estimation of the frequency of P-sample matches and E-sample correct enumerations.

College Student Scenario. To help make these issues more clear, consider the following example involving a 19-year-old college student. Assume that the student is counted in the census at his or her parents' house and also at his or her university in a different city. Also assume that the responses to the coverage probe dealing with overcounts on the census form for the parents' home does not indicate that the student may sometimes live at an alternative address (the university). If the student's name is relatively common and some of the other characteristics do not match (possibly due to nonresponse), a coverage follow-up interview may not take place, since the degree of agreement may not reach the high threshold for a coverage follow-up interview. In this case, the duplicate enumeration would remain in the census. However, if the student's name is relatively uncommon, the degree of agreement may result in a follow-up interview. In that case, the parents, when interviewed, could still assert that the student lives with them, in which case the student's duplicate enumeration would still remain in the census. However, the CCM threshold for identifying potential duplicates is almost certainly going to be lower than that for the coverage follow-up interview, which may therefore trigger an in-person follow-up interview, which might resolve the case. Assuming that the parents' home is in the CCM survey, the parents are also likely to incorrectly respond to both the CCM interviewers and the

CCM in-person follow-up interviewers as to the residence of the student. However, the accumulated evidence from the match of the student between home and university will allow headquarters staff to correctly place the student's residence at the university for purposes of CCM.

Now consider a child in joint custody. Two divorced parents each separately report the child as being resident at their households, since the child sleeps in both households about the same number of nights. Assume that the child's presence at two housing units triggers two coverage follow-up interviews (by telephone), one with each housing unit. Further assume that each parent continues to say in the interview that the child spends the majority of nights at her or his household. In this case, the child would likely remain in the census twice at two different addresses. In contrast, if both households are in the CCM, the CCM may collect the same information; however, since a person can only reside at one place, the CCM would then impute the residence probability at 0.5 at each residence, which will in the aggregate provide a superior set of counts for use in assessing the amount of net census undercoverage.

In contrast, assume in the above situation that in the coverage follow-up interview the father responds that the child sleeps at his home only two nights a week. In this case, the child becomes a duplicate enumeration in his household, which would result in the child's being removed from the census at his residence. The initial CCM interview, given the similarity in questions used, might also result in the father's including the child as a resident at his household, so the CCM will initially assert the child as resident in the father's residence. Since the child will not match to anyone in the E-sample, the situation will generate a CCM in-person follow-up interview, which may or may not correct the situation given the responses of the father to this fourth attempt at enumeration. A problem here is that one can't generally unduplicate the P-sample since it is only a sample enumeration. In situations such as this, it seems sensible to use the information from the coverage follow-up interview to resolve the situation. However, as stated above, the Census Bureau currently is not planning on doing this.

To return to the panel's concern, there is a need to retain as much data as possible to support a comprehensive analysis of the components of census coverage error to enable one to answer as many questions as possible to inform the design of the 2020 census. In particular, given the importance of the coverage follow-up interview and given that it has never been fully tested by the Census Bureau, the panel thinks it is critically important to retain as much information as possible about the functioning of the coverage follow-up interview in 2010 for a substantial sample of cases, especially for those cases in the postenumeration survey block clusters: with such information, the status of every case before and after the coverage follow-up interview is known, as well as which cases were and were not ultimately enumerated in error.

> **Recommendation 3: The Census Bureau should retain comprehensive data on the functioning of the coverage follow-up interviews for a substantial sample of cases, especially for those cases in the CCM block clusters, to support detailed follow-up analysis of the functioning of the follow-up interviews and to help suggest modifications and alternatives for use in 2020.**

CONTAMINATION

A central assumption of DSE is that census counts for PES blocks are representative of counts for all blocks in the nation. "Contamination" refers to any systematic difference in census operations for PES blocks that may affect the census counts for those blocks. The postenumeration survey interview, by taking place prior to some coverage follow-up interviews, may affect the responses received. One can imagine that some respondents may tire of the questioning and become refusals, but one can also imagine some respondents developing a greater interest in the census and improving the quality of their responses. In either case, the totality of the census will be different in the PES blocks than elsewhere, potentially biasing the estimation of the error in the census counts. To avoid contamination, CCM operations would ideally begin only after completion of regular census operations in those blocks.

However, delay of CCM operations is not without cost. Recall errors are likely to increase with the amount of time between Census Day and the beginning of PES interviews. In addition, delays in the PES will increase the number of households that involve movers, events that complicate an accurate determination of residency status. In this section, we discuss relative timing of the coverage follow-up and PES interviews with regard to the tradeoff between accuracy and contamination of the PES.

The decennial census is subject to a considerable amount of duplication, to a large extent due to the multiple ways someone can be enumerated in the census. As noted above, to reduce duplication, the Census Bureau is implementing a national search of census enumerations for duplicates, followed by follow-up interviews to confirm if the apparent duplicates are actually duplicates and to determine correct census residence. Any impact of the PES (CCM) interviews (or other PES operations) on census processes in the PES blocks is a type of contamination. One way in which contamination might operate in 2010 is if the census follow-up interviews are affected by confusion with the already completed CCM interviews. One possible effect is a refusal to participate in the coverage follow-up interviews, but it is also possible to envision more subtle impacts on the coverage follow-up interviews from CCM operations.

The impact of contamination on the entire census is essentially negligible, since the PES blocks represent less than 1 percent of the country. However, to the extent that contamination makes the CCM blocks nonrepresentative of the census in the remainder of the country, it might lead to substantial bias in the dual-systems estimates of net undercoverage.[2]

In previous censuses, waiting for the various follow-up interviews and other coverage improvement programs to be completed prior to the collection of PES data was of less concern, since these were either relatively limited operations or were concluded relatively quickly. However, the coverage follow-up interviews in 2010 may involve a large fraction of the households in the United States and take a substantial time to complete. If so, CCM interviewing may well not occur until mid-August or later, which would substantially increase the number of movers and generally reduce the quality of the data collected in the CCM. Another issue of concern is that the substantial similarity between the coverage follow-up and CCM questionnaires may cause contamination. This issue has two aspects. The first is assessing the degree to which having the CCM interviews precede the coverage follow-up interviews affects the responses collected in the coverage follow-up interviews. Attempts to measure contamination in the 1990 and 2000 censuses found no appreciable contamination (see, e.g., Bench, 2002), but, as argued above, the threat of contamination in the 2010 is more serious. If the amount of contamination is small enough to be ignored, then the Census Bureau could allow both interviews in the field in 2010; in that case, it would be valuable to assess the effects of the coverage follow-up interviews on the quality of the CCM interviews.

There were two attempts to measure contamination in the 2006 census test. The first attempt compared interview rates and rates of erroneous enumerations and omissions in the two subpopulations defined by the

[2]Previous postenumeration surveys initiated their data collection after the conclusion of the census data collection, with the minor exception of telephone A.C.E. interviewing in 2000. (In that case, there were situations where neighbors could have been in a situation where a household would still be followed up with some coverage improvement program, while their neighbor was answering the PES initial interview over the telephone.) In 2000, this meant starting the A.C.E. *nontelephone* field interviewing after the conclusion of the nonresponse follow-up and the coverage edit and coverage improvement follow-up interviewing. This was done for two reasons: (1) to avoid the possibility that the A.C.E. interview might impact the still incomplete census operations, thereby causing the PES blocks to be unrepresentative of the census, and (2) so that the evaluation that A.C.E. provided was of the complete census. However, the wait to begin the A.C.E. interviews increased the number of movers in the period between the census and the A.C.E., which reduced the quality of the data collected for A.C.E.

order of the interviews. This analysis was stratified by the various situations that result in a coverage follow-up interview (see above). However, the measurement of contamination was indirect, and the modest sample size reduced the statistical power of the analysis. There was also a matched pair design in which a second sample using the same sample design as the CCM was selected, using a block geographically proximate to the CCM sampled block. The population estimates for the two samples were compared. Again, the small sample size for this study was a concern. The results of these tests were not available prior to the drafting of this report.

The panel believes it is important to obtain more direct observation of the effects of several proximate interviews, although it is probably too late for such work in time for 2010. For example, it will be possible for a household in 2010 to be subject to four interviews. Assume that a form is mailed, with nonresponse resulting in several attempts to follow up by a field enumerator, with ultimate completion of an interview. If the specified threshold is reached, there will be attempts to carry out a coverage follow-up interview; if the household is then selected into the CCM sample, the household will be interviewed again; and, finally, if there is a difficulty in matching to the E-sample, the household could be field interviewed a fourth time. The panel is concerned that after the second or third interview, the chances either of a refusal or the collection of poor-quality data could increase. The effects may be tempered by the mix of forms of data capture, some in person and some by telephone. If this situation is likely to persist in 2020, it would be extremely important for the Census Bureau to examine, during the early part of the decade, the effects on respondents of several interviews occurring close in time, especially interviews with similar content. Such research will be complicated by the fact that households that reach the thresholds for coverage follow-up interviews are likely to be difficult-to-enumerate cases, which will confound any inference.

The second aspect of the contamination issue is what to do in 2010 if appreciable contamination is either observed or cannot be ruled out. The Census Bureau considered at least five ways to address this problem (see Kostanich and Whitford, 2005, for a discussion of some of these approaches).

One way is to combine the coverage follow-up and the CCM interviews into one multipurpose interview. However, although the interviews are similar, the CCM interview must be an independent count of a housing unit to fully satisfy the assumptions underlying DSE, and the coverage follow-up interview is dependent on information received in the initial census questionnaire. It is therefore difficult to combine these interviewing instruments.

A second approach would be to have the coverage follow-up interviews occur either before or after the CCM interviews, but apply the CCM coverage measurement program to the census before coverage follow-up interviews. This approach is referred to as evaluating a truncated census, since the definition of the census for purposes of coverage evaluation is the census that existed prior to the follow-up interviews. Any enumerations added by carrying out coverage follow-up interviews after the CCM interviews were completed could be treated as "late additions" were treated in 2000: that is, removed from the census for purposes of coverage measurement. A problem with this approach is that if the coverage follow-up interview adds an appreciable number of people or corrects the enumerations of an appreciable number of people, one is evaluating a truncated census that is substantially different from the actual census. Also, if these additions or corrections are considerably different in coverage error characteristics in comparison with the remainder of the population, it would add a bias to the dual-systems estimates. As defined, one could include the coverage follow-up interviews that occurred prior to the CCM interviews in the truncated census, in which case the net coverage error models could condition on whether a follow-up interview was carried out prior to the CCM interviews: this would remove any bias if the P-sample inclusion probabilities depended on the occurrence of the coverage follow-up interviews (but not on its outcome; for details, see Bell, 2005). Information on what the interviews added from outside the CCM blocks also could be used in these models. There are some operational complexities to this idea, including the need to duplicate the formation of relatively large processing files. Finally, as mentioned above, one is not evaluating the complete census: consequently, to assess components of census coverage error resulting from the application of the later changes from the coverage follow-up interviews, one would need to carry out a separate evaluation study outside the CCM blocks, which is a serious disadvantage.

A third approach is not to use coverage follow-up interviews in the CCM blocks. This approach avoids any contamination, but then the CCM evaluates an incomplete census, with essentially the same problems listed in the second approach, although it is worse because no results from coverage follow-up interviews could be used.

A fourth approach is to let the coverage follow-up and CCM interviews occur in whatever order they do and treat contamination in net coverage models as a constant effect times an indicator variable for which of the two interviews comes first. The difficulty with this approach is that the effect of whichever interview comes second is not clear, so it is not clear that contamination can be effectively modeled through use of a constant effect. For example, contamination might be subject to various interaction effects.

A fifth approach is to delay the CCM interviews until the coverage follow-up interviews are complete. Such a delay solves the contamination problem, but it introduces other problems. For example, coverage evaluation interviews that occurred in August 1980 were less useful than those in April due to the large number of movers that occurred during the four-month period. Thus, this approach could have a substantial, negative impact on the quality of the CCM data that are collected in 2010, depending on the length of time between the census and the CCM.

After considering these approaches, the Census Bureau decided on the last one—to delay the CCM interviews until after all coverage follow-up interviews are completed. There were several arguments given in support of this decision:

- The Census Bureau would not have to plan on having a substandard census in any area, which would certainly be true of the third approach.
- Combining the interviews, the first approach, might harm both interviews.
- The fourth approach—letting the two interviews occur whenever they fell—is speculative and would be difficult to assess prior to the 2010 census.

The second and third approaches (excluding some of the interviews) would require some assumptions about the nature of the late coverage follow-up interviews and would also require a large, parallel census database. (For details on the Census Bureau's views on contamination, see Kostanich and Whitford, 2005.)

It may be the case that further work would have demonstrated the advantages of either a truncated census (the second approach) or of combining the two interview (the first approach). Also, the panel finds some of the Bureau's arguments in relation to the second and third approach—in particular the difficulty of duplicating census processing files given the availability of inexpensive computer memory—not fully convincing. However, arguments for or against various alternatives are now moot, given the Census Bureau's decision.

The CCM interviews may not begin until late August or September 2010, which means there will be a relatively larger number of movers between Census Day and the CCM interviews in comparison with the number of movers in 2000. Data from movers in this context are known to be of poor quality, partly because a large fraction of the data collected is from proxy respondents. In addition, there also may be recall problems, since people are being queried about where they lived several months ago. This reduction in data quality will probably result in estimating fewer matches than there actually are.

An early August start for the CCM in 2010 might be possible by expediting certain operations. To determine whether this is feasible, it will be important to collect good data during the dress rehearsal, if relevant, on the possibilities of expediting the initiation of the CCM interviews and to develop a good understanding of how various delays affect the number of movers.

Given its concern over a late start to CCM interviewing, the panel would like to raise the possibility of initiating the 2010 CCM data collection prior to the completion of the coverage follow-up interviews, without any accounting for the overlap of the two data collection efforts in the estimation of net coverage error. Decisions on whether to allow these data collections to overlap, and if so, how much, are difficult to assess since they involve the comparison of two biases whose magnitudes are difficult to gauge. One bias stems from the data collected from movers, and the second bias results from the potential contamination—that the census data collected in the CCM block clusters will be different from the remaining census data. Both biases are potentially sizable and, if so, could substantially reduce the utility of the estimates from the CCM. The magnitude of these biases involves a direct tradeoff: as one moves the date for the initial capture of CCM data from mid-June until early September, the contamination bias decreases to zero as the mover bias increases substantially.

The available research does not clarify what the size of these two biases is as a function of various factors, especially the date that CCM data collection begins. The uncertainty about the magnitudes of these two biases precludes the panel from recommending how the Census Bureau should proceed. However, the panel's relatively subjective assessment of the situation is that the mover bias at its maximum (from no overlap) is likely to be substantially greater than the contamination bias at its maximum (starting CCM data collection in, say, late June). Therefore, the panel suggests that the Census Bureau reconsider starting the CCM data collection no later than mid-July, thereby allowing for some modest overlap between the coverage follow-up and the CCM data collections.

Whether or not the Census Bureau reconsiders the start date for the CCM, it should endeavor to begin CCM interviewing as soon as possible after the completion of the great majority of the census data collection, which one hopes would be before late July. Consistent with this, to the extent that it is feasible, the management of the coverage follow-up and the CCM data collections should be organized to limit the potential for contamination by selectively starting the CCM data collection in those areas in which the coverage follow-up interviewing has been completed, monitoring this on as small a geographic basis as possible. Furthermore, there are potential advantages to the use of census designs in which there is modest overlap between the coverage follow-up and the CCM, and that the Census Bureau should consider use of such designs in 2010.

Recommendation 4: The Census Bureau should organize census and coverage follow-up data collection so that data collection for the census coverage measurement (CCM) program is initiated as soon as possible after the completion of the census. In particular, the postenumeration survey in a particular area should start as soon as possible after the completion of the great majority of the census data collection—hopefully before late July. The Census Bureau should also consider census designs for 2010 in which there is some modest overlap between coverage follow-up and CCM data collections.

ADMINISTRATIVE RECORDS

The Census Bureau has explored the potential for using administrative records (data collected as a by-product of administering governmental programs) in the decennial censuses since the 1970s. Possible uses include: (1) supporting a purely administrative records census; (2) improving census nonresponse follow-up, either by using enumerator follow-up only when administrative records do not contain the required information or by completing information for households that do not respond to initial attempts by field enumerators; (3) improving the Master Address File with addresses found in administrative records;[3] (4) assisting in coverage measurement, for example, through use of triple-systems estimation;[4] and (5) assisting in coverage improvement, for example, by identifying census blocks that may not have been well enumerated or households for which the census count is likely to be in error.

One important advantage that administrative records have is that they provide a source of information for hard-to-enumerate groups that is operationally independent of the census processes. Underlying the use of a postenumeration survey is the assumption that reinterviewing people, albeit with a much more intensive interview with more highly trained

[3]The Census Bureau has already used administrative records for this purpose. The MAF is already updated using the delivery sequence file from the U.S. Postal Service, which is a type of administrative record, and the MAF is also updated using files from local jurisdictions, which are often based on local administrative sources.

[4]Triple-systems estimation is a generalization of dual-systems estimation: In this case the third system would be a merged list of individuals from administrative records (for details, see Zaslavsky and Wolfgang, 1990).

interviewers, will either generate a response when there was previously no response or will provide different information than the respondents provided earlier and thereby correct a incorrect response. One can argue that for a substantial fraction of the cases with coverage error, due to the similarity of the two requests for information, neither assumption may obtain, especially for people that are actively seeking not to be counted in the census. For many such cases, administrative records may provide the only current real chance at enumeration.

Until recently, the available administrative records have suffered from several limitations, including: insufficient coverage of the population represented on administrative records; lack of current information (particularly for addresses); lack of information on race and ethnicity; difficulty in unduplicating administrative lists with very few errors; computational burden; and concerns about public perceptions.[5] Consequently, none of the potential applications of administrative records have been implemented during a census. Until 2000, there was no comprehensive field test of the benefits of the use of administrative records for any census application, although there were assessments of the coverage of merged administrative lists in assessing the feasibility of an administrative records census. However, administrative records did support at least two major coverage improvement programs—the nonhousehold sources check in 1980 and the parolees and probationers check in 1990.

Now, however, several of the limitations just noted have been addressed. The quality and availability of national administrative records are improving, computing power has increased dramatically, and the research group on administrative records at the Census Bureau has achieved some impressive results. The primary program and database, referred to as StARS, now has an extract of a validated, merged, unduplicated residential address list with 150 million entries, 80 percent of which are geocoded to census blocks, and another extract of a validated, merged, unduplicated list of residents with demographic characteristics. These lists are approaching the completeness of coverage that might be achieved by a decennial census (Obenski and Farber, 2005).

Seven national files are merged to create StARS, with the Social Security Number Transaction File providing demographic data. As a result of this progress, an administrative records comparison will be one of six circumstances generating a coverage follow-up interview in 2010, which may be the first direct application of administrative records to assist in census enumeration.

[5]An approach to the problem of current address can be found in Stuart and Zaslavsky (2002).

However, the progress to date is not a compelling argument for widespread use. The quality of administrative records, and StARS, in support of census field enumeration is still untested, and many of the deficiencies regarding undercoverage, race and ethnicity information, and current address are still worrisome. AREX 2000 provided the only major test to date of the use of administrative records (primarily for use as an alternative method for taking a census). While the population coverage (for the more thorough of the schemes tested) was between 96 and 102 percent relative to the 2000 census counts for the five test site counties, AREX 2000 and the census counted the same number of people at the housing unit level only 51.1 percent of the time and counted within one person of the census count for only 79.4 percent of the households. So although the potential of administrative records is obvious, these ideas need further development and evaluation.

In order to make sure that an important opportunity is not being missed, but also to verify that administrative records can provide real benefits, the Census Bureau would need to support a wide-ranging and systematic research program on decennial census applications of administrative records that is amply funded and staffed. Such a program would have the specific goal of deciding which of the potential uses of administrative records are and are not feasible for use in 2020. Such decisions would have to be made by 2015 so that there would be sufficient time before the census for final testing and to best integrate these various activities into the 2020 census design. Administrative records can still be used in a limited way in the 2010 census, in addition to the role they are playing in generating the coverage follow-up interviews. In particular, administrative records might be considered either for coverage improvement or for coverage measurement in 2010. The panel believes that it no longer makes sense to view the use of administrative records as an interesting possibility for some unspecified census in the future. We believe it is crucial to comprehensively assess their potential now for use in the 2020 census. We propose six potentially feasible uses of administrative records in a census: to improve the MAF or other address lists, in late-stage nonresponse follow-up, for item imputation, to improve targeting of coverage follow-up interviews, for assistance on the status of nonmatches, and to evaluate a census coverage measurement program.

Improvement or Evaluation of the Quality of the MAF or the Address List of the Postenumeration Blocks The quality of the MAF is key to a successful mailout of the census questionnaires and nonresponse follow-up, and the quality of the independent list that is created in the CCM blocks in 2010 will be key to a successful coverage measurement program. StARS provides a list of addresses that could be used in at least two ways. First, the total number of StARS addresses for small areas could be checked against the corresponding MAF or PES totals to identify areas with large discrepancies that could be relisted. Second, more directly, address lists could be matched to identify specific addresses that are missed in either the MAF or the PES address listings, with discrepancies followed up in the field for resolution. Note that although administrative records could be used to improve the address list for either the census or the PES, to maintain independence they should not be used for both.

Assistance in Late-Stage Nonresponse Follow-Up The Census Bureau makes several attempts using field enumerators to collect information from mail nonrespondents to the census. When these attempts fail to collect information, attempts are made to locate a proxy respondent and, when that fails, hot-deck imputation (filling in for the nonresponse with the data for a randomly selected, geographically proximate household) is used to supply whatever information is needed, including the residence's vacancy status and the household's number of residents. If the quality of StARS information is found to be at least as good as that from hot-deck imputation or even proxy interviews, it might be effective to attempt to match nonrespondents to StARS before either pursuing a proxy interview or using hot-deck imputation. Especially with a short-form-only census, StARS might be sufficiently complete and accurate for this purpose. Further, one might profitably make fewer attempts at collecting nonresponse data by making use of StARS information, for example, after only one or two attempts at nonresponse follow-up, thereby substantially expediting and reducing the costs of nonresponse follow-up.

Item Imputation The Census Bureau often uses item imputation to fill in modest amounts of item nonresponse. Item nonresponse could affect the ability to match a P-sample individual to the E-sample, and missing demographic and other information may result in an individual being placed in the wrong poststratum or the use of the wrong covariate information in a logistic regression. Item imputation based on information from StARS may be preferable to hot-deck imputation. The use of StARS to provide item imputation was tested as part of the 2006 census test, but the results were not available in time for this report.

Targeting the Coverage Improvement Follow-Up Interviews The coverage improvement interview in 2010, as currently planned, will follow up households with any of the following six conditions: (1) characteristics for the additional people in large households who did not fit on the census questionnaire, (2) count discrepancies between the indicated number of residents and the number of persons for whom information is provided, (3) potential duplicates identified by a national match of census enumerations to themselves, (4) persons who, given their responses to coverage probes, may have been enumerated at other residences in addition to the one in question (potential duplicates), (5) persons who, given their responses to coverage probes, sometimes stayed at the housing unit in question and who may have been omitted from the census, and (6) people in households with different counts than in a list generated from administrative records. The workload for this operation might well exceed the Census Bureau's capacity to carry out the necessary field work given limited time and resources. It might be possible to use administrative records to help identify situations in which field resolution is not needed, for example, by indicating which of a set of duplicates is at the proper residence. (Uses of StARS in similar ways were tested in the 2006 census test, but the results were not available in time for this report.)

Determination of the Status of Nonmatches It is possible that administrative records can be used to determine the status of a nonmatch prior to follow-up of nonmatches in the postenumeration survey. It is very possible that nonmatches of the P-sample to the census may be resolved, for example, by indicating that there was a geocoding error or a misspelled name, thereby saving the expense and time of additional CCM field work.

Evaluation of the Census Coverage Measurement Program The quality of many of the steps leading to production of dual-systems estimates might be checked using administrative records. For example, administrative records information might be used to assess the quality of the address list in the P-sample blocks, to assess the quality of the matching operation, or to assess the quality of the small-area estimation of population counts. We note, however, that any operation that makes use of administrative records cannot also use the same administrative records for purposes of evaluation.

The administrative records group at the Census Bureau has already had a number of successful applications of StARS. First, an administrative records census was conducted in five counties during the 2000 census, and its quality was judged to be comparable to that of the census in those counties. (This assessment is somewhat surprising given that, as pointed out above, the agreement between StARS and the census counts was only slightly above 50 percent.) Second, StARS was used to explain 85 percent of the discrepancies between the Maryland Food Stamp Registry list of recipients and estimates from the Census Supplementary Survey in 2001 (the pilot American Community Survey).

Since the panel's suggested uses of administration records depend crucially on the quality of the merged and unduplicated lists of addresses and people in StARS, prior to the implementation of StARS for any of the above purposes in 2010 (except arguably for coverage measurement), it would be necessary to evaluate the use of administrative records in comparison to the current method used in the census. Alternatively, the use could be an additional process added to the census, in which case it would be necessary to assess the likely effects on the quality of the census enumerations along with its likely costs. If there are no opportunities for a careful test of feasibility and effectiveness of applications of administrative records in 2009, additional uses of administrative records in 2010 will not be feasible.

Thus, it is likely that additional uses of administrative records, besides their current role in coverage follow-up interviews, will have to wait until 2020. However, the 2010 census provides an important opportunity for testing the above ideas. Therefore, the panel suggests that the more promising of the above applications be developed sufficiently to support a rigorous test in 2010, with additional refinement during the intercensal period, with the goal of implementation in 2020 should the subsequent evaluation support their use. (This idea is consistent with a recommendation of the Panel on the Design of the 2010 Census Program of Evaluations and Experiments; see National Research Council, 2007.) If tests during 2010 are not feasible, the panel believes the highest priority should be given to testing during 2012–2015, as a first step toward the possible substantial use of administrative records in the 2020 census. In particular, given the promise of administrative records in relation to the census' greatest challenge, reducing omissions, we strongly advocate the testing of the use of administrative records for coverage improvement and as part of the coverage measurement program or to assess the effectiveness of the coverage measurement program in measuring the number of census omissions in 2020.

If data from StARS are used successfully in the coverage follow-up interviews in 2010 or if early tests of administrative records in the next decade strongly indicate their applicability and value for various census applications, the Census Bureau could consider even more ambitious uses of administrative data in the 2020 census. Specifically, for many housing units, the Census Bureau might use administrative data not just to replace late-stage follow-up, but as a replacement for the entire nonresponse follow-up interview. This use would seem to be especially valuable in situations in which the enumerators had determined that the nonresponding household was occupied. Under this approach, the Census Bureau would use data from administrative records to determine the occupancy status of some nonresponding housing units and the number and characteristics of its residents. To do so, the Census Bureau would have to develop criteria of adequacy of the information in the administrative records to establish the existence and count of the household for this purpose. For example, agreement of several records of acceptable currency and quality might be considered sufficient to use the information as a substitute for a census enumeration, which would reduce the burden of field follow-up.

This use of administrative records would represent a substantial change in what constitutes a census enumeration of at least the same conceptual magnitude as the change from in-person to mail enumerations as the primary census methodology. However, given that the completeness of administrative records systems and the capabilities for matching and processing administrative records has been growing, and given that public cooperation with survey field operations appears to be declining (though the mail response rate in the 2000 census was slightly better than that in 1990, reversing a trend over the past few censuses), it seems increasingly likely that administrative records will soon provide enumerations of quality at least as good as field follow-up for some housing units. Furthermore, unlike purely statistical adjustment methods, every census enumeration would correspond to a specific person for whom there is direct evidence of his or her residence and characteristics. The long-run potential for such broader contributions from administrative records is a reason to give high priority to their testing in the 2010 census.

Three possible objections might be raised in opposition to this approach. First, this use of administrative records may be ruled to be inconsistent with interpretations of what an enumeration means in the Constitution. Second, public perception that the government will otherwise obtain the information might reduce response to the census mailout questionnaire. Third, like any use of administrative records for other than their intended purpose, this may raise public concerns about a loss of confidentiality. These three issues are not compelling arguments against moving forward, but they would need to be addressed before the Census Bureau could implement their use in 2020.

In summary, if the Census Bureau is to position itself to be able to make an informed decision about the value of administrative records to fulfill a variety of possible functions in the 2020 census, it needs to make use of the various testing opportunities in both the 2010 census and in the early part of the 2010–2020 intercensal period to assess which of the applications listed here are feasible and effective. Otherwise, important benefits may be missed since one cannot implement the ideas absent a careful evaluation. Even with a successful test, there will be a number of implementation complexities that will have to be dealt with, and waiting to test such ideas in 2016 or later will likely not leave enough time.

Recommendation 5: The Census Bureau should use the various testing opportunities in both the 2010 census and in the early part of the 2010–2020 intercensal period to assess how administrative records can be used in the 2020 census.

4

Technical Issues

This chapter discusses several issues related to the proposed census coverage measurement (CCM) program for 2010: the sample design for the CCM postenumeration survey (PES), use of logistic regression models, missing data in new coverage error models, matching cases with minimal information, and demographic analysis. On several of these topics the panel offers recommendations for the Census Bureau.

SAMPLE DESIGN FOR CENSUS COVERAGE MEASUREMENT

The Census Bureau is planning a CCM PES sample of 300,000 housing units, with primary sampling units composed of block clusters (for details, see Fenstermaker, 2005). An important question concerning the census coverage measurement program in 2010 is to what extent can and should the new goal of process improvement be incorporated into the design of the CCM PES.

For purposes of CCM design, the United States will be divided into 3.7 million block clusters, and the CCM will select about 10,000 of these, each averaging roughly 30 housing units (for a total of 300,000 housing units). The Census Bureau will use an initial stratification of the 3.7 million block clusters into four types: (a) small, with between 0 and 2 housing units (as determined by the Census Bureau's Master Address File in 2009), (b) medium, with between 3 and 79 housing units, (c) large, with more than 80 housing units, and (d) block clusters of groups of American Indians on reservations. The current proposed CCM design will allocate a minimum of 1,800 housing units from about 60 medium and large block clusters per state (3,000 block clusters of the 10,000), with the remainder allocated proportionate to state population size. Also, Hawaii is allocated a minimum of 4,500 housing units in the CCM sample (roughly 150 block clusters), and 10,000 housing units (roughly 330 block clusters) are selected of American Indians living on reservations, which are allocated proportionally to the number of American Indians living on reservations in each state.

Once the 10,000 block clusters for the CCM are identified, they will be independently listed to determine how many housing units are actually present (since the MAF does not provide a perfect count and also because the MAF will be slightly dated). In particular, for small block clusters, this independent listing will find many of them to have more than two housing units. If the number of housing units for small block clusters is found to be more than 10, current plans are to choose those block clusters into the CCM sample with certainty. Otherwise, the remaining small block clusters will be subsampled. (Plans are to subsample small block clusters with between none and two housing units at the rate of 0.1, those with between three and five housing at the rate of 0.25, and those with between six and nine housing units at the rate of 0.45.)

Finally, regarding substate allocations of block clusters, while plans are currently not final, the Census Bureau is likely to include some modest degree of oversampling of block clusters in areas that have a large fraction of people that rent their residences and possibly in areas that have a large fraction of minority residents.

The general argument in support of the state allocations for the 2010 CCM PES is that they mimic those for 2000, since the Census Bureau was generally satisfied with the 2000 design of the Accuracy and Coverage Evaluation (A.C.E.) Program in terms of the variance of estimates of net undercoverage for poststrata. (The Census Bureau has no specific variance requirements for the 2010 CCM estimates, because production of adjusted counts is not anticipated.) With respect to substate allocations, the Census Bureau is concerned with increased variances and so intends to refrain from more than a modest amount of oversampling.

The Census Bureau examined some alternative specifications for the design of the CCM PES to see if they might have advantages, using simulation studies of both the quality of the resulting net coverage error estimates and the quality of estimates of the number of omissions and erroneous enumerations at the national level and for 64 poststrata (for details, see Fenstermaker, 2005, 2006). Initially, four designs were examined: (1) the design described above—i.e., allocations proportional to total state population, with a minimum of 60 block clusters per state, with Hawaii allocated at least 150 block clusters; (2) as (1) except with

Hawaii allocated at least 60 block clusters; (3) allocations to the four census regions to minimize the variance of estimates of erroneous enumerations, and within regions, allocations are proportional to state size; and (4) half of the sample is allocated proportional to the number of housing units within update/leave areas[1] and half is allocated proportional to each state's number of housing units. Through use of simulations, for each design and resulting set of PES samples across simulation replications, national estimates were computed of the rate of erroneous enumerations (and the rates of erroneous enumerations from mail returns, from nonresponse follow-up, and from coverage follow-up), the nonmatch rate, the omission rate, and the net error rate. National estimates of the population were also computed, along with their standard errors. The same analysis was done at the poststratum level. One hundred replications were used for the simulation study. The results supported retention of the design that closely approximated the 2000 A.C.E. design (described above). A subsequent analysis added an additional six proposed sample designs for analysis.

The panel supports the overall sample size of 300,000 housing units, which was also endorsed, as part of Recommendation 6.1, by the Panel to Review the 2000 Census (National Research Council, 2004b). Such a design would produce net coverage estimates of similar precision to those of the 2000 A.C.E. The adequacy of the CCM sample size is somewhat supported by the adequacy of the A.C.E. sample size, though the objectives of these surveys have changed and therefore arguments used to support the A.C.E. sample size may no longer be fully relevant. However, such a position is necessary given the current lack of experience estimating the components of census coverage error.

Aside from sample size, the selection of a sample design for the CCM in 2010 will involve addressing related but somewhat competing goals, given that there are two overall objectives of the coverage measurement program for 2010. First, there is the primary objective—the measurement and analytic study of components of census coverage errors. Second, there is still the need to be able to measure net coverage error for at least three reasons: (1) to estimate the number of census omissions, (2) to serve the many users who remain interested in assessments of net error at least for states and major demographic groups, and (3) to facilitate comparison with the quality of the 2000 census.

To address the first general goal, one would like to target problematic domains—determined using 2000 census data or data from the American Community Survey (ACS) for which there is predicted to be a high fre-

[1] These are areas in which the enumerator updates the address list, and, at the same time, drops off a questionnaire to be filled out and returned by mail.

quency of various types of census coverage error. (In an optimal design for any individual component, the sampling rate would be proportional to the stratum standard deviation, which is likely to be higher in strata where the particular coverage error is greater.) However, one has to be careful because one also has to have a facility for discovering any unanticipated problems that might appear in areas that were relatively easy to count in 2000. Each census seems to raise relatively novel sources of census coverage error, and at the same time, each census seems to have areas that were hard to count a decade earlier and subsequently are relatively easy to count.

Yet the goal of producing high-quality estimates of net coverage error for all states and for all major demographic groups calls for a design that is somewhat less targeted. As with estimation of components of error, the most efficient design for estimation of net coverage error would over-sample areas with high rates of omissions or erroneous enumerations. This then allows reducing the sampling weights associated with individual blocks expected to exhibit the most variance in the two components of net error. However, it is especially critical for net error estimation to avoid extreme undersampling in any areas because large sampling weights will quickly inflate variances if associated with blocks having more problems than anticipated.

Another way to look at the situation is that there is a modest tension between the need for cross-U.S. reports on net coverage error, and the need for specific analytic studies on possibly problematic processes. So if one had a list of potentially worrisome places where census processes are likely to enumerate certain kinds of housing units with a high frequency of coverage error, those places should be oversampled in the 2010 CCM design. But this should be done while maintaining the ability to produce reliable estimates of net coverage error at some level of geographic and demographic detail.

Given this modest tension, the panel believes that the Census Bureau has selected a design that may not sufficiently accommodate the primary goal of measuring and analyzing components of census coverage error. The state allocations of the Census Bureau's proposed CCM sample design are too oriented toward producing state-level estimates of net undercoverage of comparable quality with the 2000 estimates. Instead, the new purpose of CCM in 2010 should be accommodated by modifying the state allocations of block clusters to include more block clusters from states that are predicted to be harder to count, by including a greater degree of oversampling of substate areas that are likely to be difficult to count (though the latter is clearly dependent on the Census Bureau's as yet unspecified plans), or both.

The analysis carried out by the Census Bureau of 10 sample designs for state allocations is thorough. However, with respect to substate allocations of block clusters, the Census Bureau might consider, in addition to oversampling medium and large block clusters with a high percentage of renters, oversampling block clusters with large percentages of individuals or housing units with other features that might be associated with census coverage error, such as: (1) small multiunit structures, (2) a high percentage of foreign-born residents in 2000, (3) a high percentage of proxy interviews in 2000, (4) a high percentage of whole household imputation in 2000, (5) a high percentage of vacation homes, or (6) recent additions to the housing stock. In addition, as in 2000, the Census Bureau could oversample blocks in which there is a higher chance of geocoding errors or areas in which there was a high percentage of additions through the Local Update of Census Address (LUCA) Program or block canvass adds or deletes.[2] It is likely that efforts devoted to modifying substate allocations will be more important than the state allocations, but both deserve attention.

In addition to the above general suggestions, the panel has a specific suggestion for the 2010 CCM sample design that provides a reasonable compromise between designs that are focused on estimation of net coverage error and designs that focus on components of census coverage error. We urge the Census Bureau to evaluate a CCM sample design that retains the identical structure of the current census design for a substantial fraction of the sampled units, possibly 60 to 75 percent (by making the obvious change to the sampling rates) and allocates the remaining sample to anticipated problematic regions or block clusters. Such a change would potentially provide a much greater number of census coverage errors to support models examining which factors relate to coverage error. At the same time, allocating the bulk of the sample to a general purpose design would limit the risk of inflated variances for net error estimates associated with finding large errors in unexpected locations. Research on what percentage to use and how this compares with the Census Bureau's proposed design can be carried out using simulation studies of the type the Census Bureau has already carried out, though it also would be very useful to incorporate some accounting for any differences that are expected to be seen between 2000 and 2010 (possibly based on the ACS).

In conducting additional simulations, we propose that the Census

[2]In considering characteristics that can serve as the basis for oversampling, it is important to stress that even if some problematic circumstances are identified, it will generally be the case that very little individual household-level information could be used as the basis for oversampling since such information would have to be available on the MAF. However, area-wide frequencies of the same characteristics can often provide reasonable surrogates. For example, areas with many renters can be targeted, but one cannot target renters individually for oversampling.

Bureau also reconsider the metrics it uses to compare and assess 2010 CCM sample designs. In its simulations, the Census Bureau examined estimates of the coefficient of variation of estimates of net error, rate of erroneous enumerations, rate of omissions, and the rate of P-sample nonmatches. The Bureau also looked at coefficients of variation for net error estimates for groups of poststrata from 2000. The panel would like to suggest, in addition, the use of several additional types of metrics. Letting DSE_i = the direct dual-systems estimate for an aggregate i (e.g., state by major demographic group), E_i = the E-sample total for an aggregate i, P_i = the P-sample total for an aggregate i, I_i = the number of imputations for an aggregate i, EE_i = the number of erroneous enumerations for an aggregate i, and M_i = the number of matches for an aggregate i, the panel believes that the following metrics would provide more direct indication of the benefits of alternative CCM designs:

$$\left(\frac{DSE_i + I_i}{E_i + I_i}\right), \frac{EE_i}{(E_i + I_i)}, \text{ and } \frac{I_i}{(E_i + I_i)}$$

The first metric is intended to be evaluated at the block cluster level (based on synthetic estimation), while the remaining two metrics are computed at the state level. The first metric,

$$\left(\frac{DSE_i + I_i}{E_i + I_i}\right)$$

is a local undercount estimate since $DSE_i + I_i$ is similar to the dual-systems estimate and $E_i + I_i$ is an estimate of the census count. The second metric,

$$\frac{EE_i}{(E_i + I_i)}$$

is a measure of the percent of erroneous enumerations. The third metric,

$$\frac{I_i}{(E_i + I_i)}$$

is a measure of the percent of whole-person imputations. The last two metrics therefore assess the degree to which an area is encountering problems in enumeration. The goal then is to select a CCM sample design that produces estimates of these quantities at the indicated level that have substantially lower variances than those from the currently proposed design.

Simulation studies of the design alternatives mentioned above, using these metrics, may identify designs that are nearly as effective as the Census Bureau's current design at estimating net coverage error at the level of states and major demographic groups while increasing the number of sampled census coverage errors. The panel believes there still may be sufficient time to carry out this analysis.

The panel's suggestion that the Census Bureau consider additional oversampling of difficult-to-count housing units in the 2010 CCM design is incomplete without considering what data should be used in support of this effort. Certainly, the Census Bureau could continue to use census and A.C.E. data from 2000, as in the above simulations, possibly making some effort to better identify erroneous enumerations and omissions given the weakness of A.C.E. data for that purpose. However, that might be too time-intensive an activity as 2010 nears. Census, ACS, and StARS (see Chapter 3) data could also be used as the basis for an artificial population study, in which the components of census coverage error were "imposed" on the census enumerations. That is, statistical models using current best guesses for causal factors and their impact on coverage error could be developed, relating person, household, and contextual characteristics to probabilities of duplication, omission, and being counted in the wrong location. Then, a number of simulated censuses and PESs could be conducted, with people and housing units missed, duplicated, and counted in the wrong place with various probabilities. Erroneous enumerations, as defined here (which excludes duplications and enumerations in the wrong location), would be more difficult to incorporate in such a study since one does not have a base population to apply a model to. However, this component is likely the least important to address, and there may not be an effective causal model predicting which newborns are erroneously included in the census, which recent deaths are erroneously included, which visitors are included, and which fictitious individuals are included. If the suggested study is carried out, then, analyses in 2010 to identify which factors are and are not associated with various components of coverage error can be used to refine the models used for incorporating components of coverage error to better plan the coverage measurement data collection in 2020.

Finally, a very serious complication in carrying out this research plan is that many of the most important predictive factors in statistical models of components of census coverage error will have to be indicator variables for the various census processes used in association with the enumeration of each housing unit or individual. (This requirement results from having a feedback loop that identifies census processes in need of modification.) However, the census processes are generally not represented on the standard census files or in A.C.E. in 2000. This lack strongly argues for the collection of a master trace sample (a sample of households for which the entire census procedural history is retained) in 2010 and that the designs of the CCM sample and the master trace sample be such that there is substantial overlap between them. For current work, and in the case that a master trace sample database is not constructed in 2010, the Census Bureau should determine how it can use census management information files to populate an analysis database that represents the components of census coverage error and as many as possible of the predictors discussed in Chapter 5.

None of the approaches suggested here as to how to examine the optimal extent of oversampling problematic households is ideal, which is not surprising. The Census Bureau does not have good historic information on how coverage errors are related to census processes, which makes targeting the sample much harder and which makes simulating the situation harder as well. However, the Census Bureau has acquired a lot of information about the circumstances that cause some of the coverage errors and where those more problematic areas are located; those areas need to be oversampled to some extent. The coverage problems do change from census to census and some of the problematic areas are due to idiosyncrasies that appear for only a single census. Yet it is sensible to assume that much of the causal nature of coverage error is persistent across at least a few censuses, and that is what needs to be captured in the CCM survey. So focusing on areas with high proxy interview rates or high imputation rates in the previous census, on areas with a large percentage of vacation homes, or on areas with many small, multiunit housing units (though this has some difficult definitional and implementation aspects) is likely to be beneficial in the design of the 2010 CCM survey since such households have been and are likely to remain hard to count. Of course, over time, new problems will crop up, and old ones will be addressed, and so the process of census improvement will be a dynamic one.

Given that the design of the 2010 CCM PES needs to target block groups that have a higher frequency of housing units that are vulnerable to census coverage error, the Census Bureau should give serious consideration to alternative designs that, without sacrificing much efficiency in estimating net coverage error, could provide a larger number of (anticipated to be) hard-to-enumerate households and individuals. Such a design would improve the estimation of parameters of the statistical models linking coverage error to census procedures. In particular, the Census Bureau should consider implementing a design that mixes a high proportion of cases selected using the current design with a smaller proportion of cases in hard-to-enumerate areas. This design could be assessed through simulation studies like those the Bureau has already used, supplemented by additional metrics suggested here.

Recommendation 6: The Census Bureau should compare its sample design for the 2010 census coverage measurement postenumeration survey with alternative designs that give greater sampling prob-

ability to housing units that are anticipated to be hard to enumerate. If an alternative design proves preferable for the joint goals of estimating component coverage error and net coverage error estimation, such a design should be used in place of the current sample design.

LOGISTIC REGRESSION MODELS

In the last few years the Census Bureau has devoted a considerable amount of its resources on coverage measurement research to improving the estimation of net coverage error in 2010, with a primary focus on developing two logistic regression models to replace poststratification to address correlation bias. Any small-area estimates of net coverage error will likely be based on these same logistic regression models, replacing the use of (so-called) synthetic estimation. Both poststrata and synthetic estimation were used in the coverage measurement programs in 1990 and in 2000, so the current plan is a substantial change to the estimation of net coverage error at the level of both large and small domains. Despite the new focus on estimating components of error, there remains good reasons for devoting considerable attention to the estimation of net error.

First, given the focus in 2000 on net error estimation, the data available from A.C.E. are not directly useful as substitutes for the data on components of coverage error that will be collected in 2010. An important example of this is the different definitions of correct enumerations in 2000 and 2010, which suggests that the frequency of erroneous enumerations and omissions will likely be substantially less and of a somewhat different nature in 2010 than they were in 2000. As a result, any attempts to model the 2000 A.C.E. data without accounting for various differences between 2000 and 2010 will probably provide limited guidance for how to estimate components of coverage error in 2010. (However, we believe that some efforts in this direction are warranted.)

Second, as argued in Mulry and Kostanich (2006), estimating net coverage error facilitates estimation of the number of census omissions. Therefore, some focus on estimation of net coverage error is justified.[3] Third, as noted in Chapter 2, strong interest remains for many census data users in the assessment of net coverage error, in particular for demographic groups, but also for states and cities.

---

[3]Although the Census Bureau will use estimates of net coverage error to develop estimates of the number of census omissions for domains that will support various tabulations, we hope that the Census Bureau will develop analytical models based on the P-sample individuals that are determined to be census omissions. The main disadvantage of doing this is that this analysis may miss the types of census omissions that are not captured in either the census or the P-sample, which are collectively estimated using the Census Bureau's approach.

The Census Bureau plans to use logistic regression for fitting the probability of match status for the P-sample and the probability of correct enumeration status for the E-sample. Logistic regression is more flexible than poststratification in terms of handling continuous predictor variables and selective use of interactions among predictor variables. This flexibility potentially allows inclusion of more predictor variables without increasing the variance of estimated probabilities. Furthermore, logistic regression is a model that, in this context, is applied at the level of the individual; therefore, information collected at that level can be easily used in conjunction with information that is collected at a more aggregate level. Finally, not only is logistic regression likely to be better than poststratification in estimating net coverage error for these reasons, but it is also much better suited for the analytic purposes of providing a better understanding of which factors are and are not related to net coverage error than poststratification.

Poststratification is mentioned in the earliest literature advocating the use of dual-systems estimation (DSE) to measure populations (Sekar and Deming, 1949), and it has been used in the census since the 1980 postenumeration program to reduce correlation bias. Poststratification simply means that one partitions the CCM sample data into groups that are more homogeneous and then separately estimates the adjusted population counts

$$(C - II)\left(\frac{CE}{E}\right)\left(\frac{P}{M}\right),$$

within those poststrata.[4] A perfect poststratification would partition the P-sample population and the E-sample population so that the underlying enumeration propensities for individuals in a poststratum were identical. However, this is unattainable and therefore the practical goal is to partition the sample cases so that individuals are more alike within a poststratum than individuals are from different poststrata. If this is accomplished, correlation bias should be reduced (see Kadane et al., 1999, for details).

Poststratification also supports the use of synthetic estimation, which carries down adjustments to census counts to low geographic levels. Synthetic estimation makes use of coverage correction factors,

$$\left(\frac{C - II}{C}\right)\left(\frac{CE}{E}\right)\left(\frac{P}{M}\right),$$

[4]See Chapter 3 for definitions. Note that *CE* is defined consistent with the definition of a correct enumeration in A.C.E., that is, an enumeration that is located in the search area.

which are applied to any subpopulation in a poststratum by multiplying the appropriate factor by the relevant subpopulation's census count to produce the adjusted count for that subpopulation. To produce geographic estimates, which often requires adding subpopulations that belong to different poststrata, one simply sums the associated adjusted counts.

Estimates of the variance of synthetic estimates for small domains are necessarily a combination of estimates of the variance of the coverage correction factors for the poststrata involved (depending on the domains) and a residual variation due to any unmodeled heterogeneity of the relevant subpopulations of interest within the required poststrata. The first component can be estimated by standard methods. However, estimation of the second variance component is more difficult.

As mentioned above, although poststratification has the advantages of reducing correlation bias and supporting synthetic estimation, a major disadvantage is that, as applied by the Census Bureau, it allows only a relatively small number of factors to be included in the poststratification scheme (and in the resulting synthetic estimation). This limitation exists because the Census Bureau typically includes the full cross-classification of the factors used to define the poststrata, and, as a result, the individual poststrata quickly become very sparsely populated, despite the large sample size of the PES. Use of more poststratification factors, and therefore more poststrata, trades off greater homogeneity in each poststratum at the price of higher sampling variances for the coverage correction factors. Furthermore, the fact that the various poststrata generally share some characteristics with other poststrata (for instance, there are many poststrata for Hispanic women) is generally ignored in the associated estimation. As a result, there is a failure to pool information when it may be beneficial to do so.

The alternative that is being planned for use by the Census Bureau in the 2010 CCM is logistic regression of both the binary match/nonmatch variable and the binary correct enumeration/not correct enumeration variable. Poststratification is a special case of logistic regression in this context in which the predictors of the logistic regression are indicator variables for membership in the categories defining the poststrata, and all interactions are included in the model. In theory, for the same reasons that logistic regression may be preferred to poststratification at the aggregate at which that analysis is carried out, small-area estimates that are based on the probabilities of match and correct enumeration status estimated using logistic regression could improve on those provided through synthetic estimation by effectively averaging over more of the data.

In the following, a number of issues relevant to the use of logistic regression are raised and discussed, and a variety of suggestions are given. In addition, it should be noted that Chapter 5 contains Recommendation 10, which provides a list of key issues that need to be addressed by the Census Bureau in going forward with research on this topic.

Logistic regression was first suggested for use in the general area of DSE by Huggins (1989) and Alho (1990) and specifically applied to census undercoverage in Alho et al. (1993). However, these studies did not consider how to treat cases with unresolved match status or unresolved correct enumeration status, and they made use of the data only from the P-sample blocks, rather than the full census. Haberman et al. (1998) introduced some additional features that addressed the above limitations. They proposed two separate logistic regressions to model match status (using P-sample data) and correct enumeration status (using the E-sample data). To represent cases with unresolved match status (with a completely analogous discussion of correct enumeration status), two records are constructed, one "matched" and the other "unmatched," and weights are used to represent the "probability" that a given record matches to the census, given the available characteristics. (Match and correct enumeration probabilities for unresolved cases could be provided by a computer matcher developed by the Census Bureau.) Survey weights are also attached to all the records to reflect the complex sample design. This approach is the Census Bureau's leading candidate to support net coverage error modeling in 2010.

One can see how these two logistic regression models relate to DSE as follows. In the formula for DSE,

$$(C - II)\left(\frac{CE}{E}\right)\left(\frac{P}{M}\right),$$

the probability that a census enumeration is correct is the second factor, and the probability of a match is the inverse of the third factor. Therefore, the two logistic regression models estimate two of the three main factors in DSE, with the remaining factor being the number of matchable enumerations in the census (and is directly measured).

Using logistic regression, synthetic estimation can now be replaced by the following methodology. Letting $\hat{p}_{CEi}$ represent the estimate from logistic regression of the correct enumeration probability for person i and letting $\hat{p}_{Mi}$ represent the estimate from logistic regression of the match probability for person i, the estimated number of people in a small area is the sum of the ratio $\hat{p}_{CEi} / \hat{p}_{Mi}$ over the individuals i in that area (ignoring the treatment of cases with insufficient information for matching).[5] A grouped jackknife procedure is used to obtain the standard errors of the

[5]In this discussion, we are ignoring missing data in covariates, which introduce some complexities.

small-area estimates. If the explanatory variables are limited to those collected in the census and are not characteristics or process variables from CCM, small-area estimates for any subdomain can be computed directly using the method just described. However, as noted above, this approach sacrifices the additional predictive power of covariates collected for cases in the P-sample. Techniques suggested by Eli Marks may be used to accommodate the use of P-sample variables at the subnational level (for details, see Marks et al., 1974).[6]

There are some complications in using this approach to small-area estimation. One issue is how to incorporate the survey weights in the model-building and model-fitting processes. The CCM PES sampling weights need to be incorporated not only in the estimation of the logistic regression coefficients, but also in the decision as to which predictors to include in the logistic regression models and which model form to use, as well as in estimating the variance of the resulting estimates. The question of how to treat the complex sample design in these types of models has a substantial research literature. The approach taken by the Census Bureau is to weight the cases using the sampling weights. An alternative approach, which may produce more efficient estimates, is to include the variables that make up the sampling weights as predictors in the model (see, e.g., Little, 2003). Comparisons of these two approaches would be of interest.[7]

A second complication is the treatment of missing data. Specifically, it is not clear how to effectively treat cases with insufficient information for matching in the estimation of the relevant logistic regression coefficients. Regarding the small-area estimation that results from the use of the logistic regression model, it is also not clear how to treat non-data-defined

[6]The Census Bureau has examined competing estimators that all have empirical deficiencies in comparison to the above estimate. As mentioned, the estimate for the population of a domain is $\sum_i \hat{p}_{CEi} / \hat{p}_{Mi}$ for all individuals i in a domain. A competing estimator that the Census Bureau has mentioned is $\sum_j w_j \hat{p}_{CEj} / \hat{p}_{Mj}$, which is now a sum over the individuals j in the PES blocks and in the relevant domain. Another competing estimator replaces the correct enumeration probability, $\hat{p}_{CEj}$, in these two alternatives by an indicator function for those individuals in the domain that had correct enumeration status, reducing the modeling to only the logistic regression model of match status. The problem with these two alternatives is that they are too sensitive to sampling variation. The Census Bureau has also considered variants of these two alternatives by reweighting the data elements so that the data-defined people in the E-sample are ratio adjusted to the census counts in poststrata. A further problem with some of these approaches is that small-area estimates do not necessarily sum to the estimates for larger areas.

[7]Another complication which we do not discuss is that the adjustments made on the A.C.E. research file have resulted in the dependent variables occasionally lying outside of the interval (0,1).

cases in the census. To address this complication, we provide some guidance below in the section on missing data issues in coverage modeling.

The Census Bureau has focused much of its efforts to date regarding developing the logistic regression approach on the performance of six models for both the P-sample matches and the E-sample correct enumerations. These logistic regression models all use explanatory variables that are indicator variables of various combinations of the levels of six factors used to define the 416 poststrata used in the March 2001 net undercoverage estimates: race/origin (seven groups), age/sex (seven groups), tenure (owner, nonowner), metropolitan statistical area/type of enumeration area[8] (MSA/TEA) (four groups), region (four groups), and mail return rate (high or low, with boundaries dependent on race/origin domain). There are six sets of explanatory variables:

1. The 416 indicator variables for the poststrata used in the March 2001 poststratification.
2. The 150 main effects and first-order interactions of the variables used to define the March 2001 poststratification.
3. The 23 main effects of the variables used to define the March 2001 poststratification.
4. The 98 main effects and all interactions from the variables for three of the six factors from the March 2001 poststratification—race/origin, age/sex, and tenure. The acronym ROAST (race/origin, age/sex, tenure) is used to distinguish this reduced set of factors from the full set used in the 2001 poststratification.
5. The 62 main effects and first order interactions from ROAST.
6. The 14 main effects from ROAST.[9]

These six models were fit to data from the A.C.E. research database (for further details, see Griffin, Mule, and Olson, 2005).

The Census Bureau was interested in comparing the five logistic regression models (models 2–6) against the 2001 poststratification (model 1) with respect to the ability to fit A.C.E. data in a predictive setting. A predictive setting is appropriate since the models are fit using PES data but are then applied to the entire census data set for small-area estimation. Certainly, if any of the models 2–6 were evaluated to be equivalent

[8]Type of enumeration area indicates which of the three main types of initial enumeration was used: (1) mailout–mailback, (2) list–enumerate, or (3) update–leave.

[9]The number of interactions does not correspond to the situation of fully crossed effects since the poststratification used in 2000 did not fully cross the six variables. For example, the poststrata of American Indians or Alaskan Natives living on a reservation is only crossed by age/sex and tenure, but not by MSA/TEA, region, or mail return rate, and this extends to the ROAST models.

to model 1 in such a comparison, that would argue for use of logistic regression in replacement of poststratification, since logistic regression can make use of more variables, especially continuous variables, in addition to those used to define the poststrata. A more appropriate comparison would seem to be between the 2000 poststratification and logistic regression models that allow the use of additional variables chosen to provide additional predictive power.

When comparing nested models—that is, models that are identical except that some of the parameters in one of the models have been constrained to be constant (typically zero, which is equivalent to removing the associated predictor from the model)—the distinction between fitting and prediction is typically represented by adding a penalty factor to a goodness-of-fit measure for including additional predictors in the larger model. As in linear regression, the additional parameters guarantee that the model with the larger set of predictors will fit the data at least as well as, if not better than, the more parsimonious model, but this advantage may be offset by the increased variances of the fitted values, due to the estimation of more parameters. The combination of the goodness-of-fit statistic and the penalty for additional parameters reflects this tradeoff. Haberman et al. (1998) suggested using a logarithmic penalty function to address this issue, with a jackknife bias estimate to adjust for the use of unnecessary predictors (overfitting). Measures such as Mallows' C_p (Mallows, 1973) and the information criteria AIC and BIC also provide useful penalties for comparing regression models in a predictive situation.

The situation for comparing nonnested models is less straightforward but important to address since the Census Bureau may need to make such comparisons. For example, one such nonnested alternative put forward by the Census Bureau separates the modeling of undercoverage into two models, one for the probability that an entire household will be missed and another for the probability that an individual in a partially enumerated household will be missed (for details, see Griffin, 2005). The panel agrees with the Census Bureau that cross-validation would be a suitable technique for comparing rival nested and non-nested models.

In cross-validation, the sample is split so that the model can be fitted to one part, and the accuracy of predictions evaluated on the other part; the accuracy of prediction is thus not overstated due to fitting and evaluating the model on the same data. A standard approach is to split the data into several equal-sized pieces and remove each piece in turn from the fitting data set. The performance of each fitted model is assessed using some loss function in predicting the values for the set-aside fraction, and the loss function is averaged over all of the replications so defined.

The Census Bureau implemented cross-validation using 100 equal-sized groups, and the loss function used was the logarithmic penalty function from Haberman et al. (1998). Finally, the average over all 100 groups was weighted using the A.C.E. survey weights.

The results of the Census Bureau's cross-validation comparison of the five alternative logistic regression models to the 2000 A.C.E. poststratification (Griffin, Mule, and Olson, 2005) are given in Table 4-1. The Correct Enumeration column provides the cross-validation statistic for each of the six models in estimating the correct enumeration rate, and the Match column provides the cross-validation statistic for each of the six models in estimating the match rate.[10] The similarity of fit across the models suggests that many of the interactions in the poststratification model are relatively small. These findings also support the view that even the most effective of the five alternative models, model 2, offers only minor benefits over the full poststratification. However, it should be noted that these models are limited to the use of the variables in the 2000 poststratification and do not assess the potential of other predictors or model forms.

The panel also used cross-validation to assess the impact of the use of survey weights on the performance of the model (following DuMouchel and Duncan, 1983). Using the logistic regression model with only the main effects from the poststratification (model 3), we formed 100 groups for the cross-validation. (This was done in two ways to examine the degree to which the block clusters were homogeneous. In one computation, we randomly selected P- and E-sample people into 100 groups for cross-validation without regard for block cluster membership; in the second computation, we randomly selected P- and E-sample people into 100 groups while maintaining the block cluster structure of the A.C.E. sample design.) Using the cross-validation, we compared the performance of the logistic regression model unweighted by the survey weights with the performance of the logistic regression model weighted by the survey weights. We assessed performance using a weighted log likelihood penalty function. The results are given in Table 4-2. The results suggest that use of the survey weights in computing the logistic regression coefficients substantially improves performance in comparison with unweighted fitting, as assessed by the weighted criterion. This result raises the possibility that inclusion of the survey design variables as predictors may provide a distinct improvement in these models.

Recently, the Census Bureau (Mule et al., 2007), motivated by nonlinear residual-type plots, substituted a spline function for the (essentially)

---

[10]The ranks observed of the average weighted log likelihoods across models—and to a substantial degree even the average weighted log likelihoods themselves (not shown here)—did not change when the number of cross-validation replications changed from 100 to 25 or 20 (also not shown here).

TABLE 4-1 Cross-Validation of Six Preliminary Logistic Regression Models: Average Weighted Log Penalty Function

Model	No. of Parameters	Correct Enumeration	Match
1. Poststratification	416	.2351	.2603
2. Main Effects and Two-Way Interactions	150	.2349	.2598
3. Main Effects	23	.2354	.2598
4. ROAST	98	.2355	.2617
5. ROAST Main Effects and Two-Way Interactions	62	.2355	.2618
6. ROAST Main Effects	14	.2360	.2619

NOTE: The logarithmic penalty function that was used in the cross-validation for the correct enumeration rate modeling was

$$-\frac{1}{W_E}\left\{\sum_{i\in E-sample} w_{ei}\left[p_{CEi}\log\left(\hat{p}_{CEi}\right)+\left(1-p_{CEi}\right)\log\left(1-\hat{p}_{CEi}\right)\right]\right\},$$

where W_E is the weighted total for the E-sample, w_{ei} is the sampling weight for the jth E-sample individual, p_{CEi} is the correct enumeration status, and $\hat{p}_{CEi}$ is the predicted correct enumeration status from the model. An analogous function was used for modeling the match status. Given the negative sign in this expression, smaller values of this statistic imply a better fit to the data.

TABLE 4-2 Cross-Validation Assessment of the Effects of Survey Weights

Weighted/ Unweighted	Type of Selection	Average Log-Likelihood Penalty Function Over 100 Cross-Validated Replications: E-sample	P-sample
Unweighted			
	Random selection	.2782	.3278
	Maintain clusters	.2785	.3281
Weighted			
	Random selection	.2357	.2615
	Maintain clusters	.2360	.2619

NOTE: See note to Table 5-1 for details on the average log-likelihood penalty function.

four indicator variables for age ranges, which provided a piecewise linear and quadratic function that was selected to fit the observed relationships between age and both match rate and correct enumeration rate. This substitution would not have been available using poststratification. Initial indications are that this substitution provides only modest benefits for the overall fit of the logistic regression models, but there may be substantial advantages for estimation of specific demographic groups, particularly age groups that are not well-estimated by the four-part step function for age, which is roughly people aged 17–21. The panel strongly supports this research. Other continuous variables that might be productive are contextual variables that are estimated at the area level, such as percent vacant, percent renter, and percent minority.

Along these lines, the panel strongly suggests moving away from the predictors that are identical to those used in the previous poststratification to more fully exploit the flexibility of logistic regression. However, as mentioned in the discussion concerning use of logistic regression models to substitute for synthetic estimation, any predictors used in these logistic regression models must be available from the census to support estimation of net census error for any domain (at least in the form currently proposed for use by the Census Bureau). This requirement restricts the available predictors to functions of the six factors used to define the A.C.E. poststratification, a few additional variables from the short form in 2010, any variables collected during census processing, and contextual variables collected at aggregate geographic levels (say, from the American Community Survey or StARS). Schindler (2006) examined many of these possible variables to assess whether they provided substantial additional benefits as additional factors in producing poststrata (but not in a logistic regression approach). He considered the following variables: (1) geographic—census region, state, urban–rural, and mode of census data collection (mailout–mailback, list–enumerate, or list–leave); (2) contextual variables at the tract level—mail return rate, and percentage minority; (3) family and housing variables—marital status, relation to the head of household, and structure code (single unit or multiunit); and (4) census operational variables—indicator of mail or enumerator return, date of return, and proxy status. This research did not identify any variables that provided substantial benefit over the 416 indicator variables from the poststratification used in A.C.E.

This analysis, while extremely important, should not be considered conclusive. For example, in the related problem of examining large numbers of subsets of a collection of possible predictors for use in regression-type models, it is difficult to know whether one has missed an effective combination of variables (or transformed versions, or interactions, etc.). This is complicated work, and it may be that further examination of potential predictors may still prove useful. The panel notes that to assess the novel contributions of sets of covariates that are highly correlated, principal component analysis might provide useful insights.

As with all statistical models used in predictive settings, the explanatory variables used should, to the extent possible, be consistent with what is known about the factors that are related to census coverage error. There have been a variety of studies, especially ethnographic work, that inform understanding about why certain housing units are missed in the census and why people with various characteristics are missed in otherwise enumerated housing units. (A good bibliography is provided by Jaros, 2007.) This information is moderately consistent with the variables currently included in the logistic regression models being examined by the Census Bureau, but the linkage between the research findings and the predictors in these models is not as direct as one would like. The logistic regression models should reflect what is known about the sources of census coverage error, to the extent that this information is represented on the short form and in available contextual information.

The panel believes that the Census Bureau's initial focus on the covariates previously used in the 2000 poststratification, rather than a broader look at the variables available, was due to the desire to determine relatively quickly whether such a technique was an improvement over poststratification and could be relied on in a production environment. This was important to determine, but the use of additional covariates is not a severe complication, and no major additional software development would be required. There may therefore have been an unnecessary focus on evaluating a model very similar to the poststratification used in 2000. We hope that there is still time to operate in a more exploratory manner prior to 2010. The six models that have garnered the majority of attention to date are too similar to each other to learn enough about what collections of predictors will work well. The Census Bureau should therefore expand the important research carried out by Schindler (2006) and apply it to the logistic regression models, attempting to identify additional useful predictors for match rate and correct enumeration rate, using cross-validation to evaluate the resulting logistic regression models.

Another issue is whether the same predictors are planned for use in both the logistic regression model for match rate and correct enumeration rate in order to eliminate so-called balancing error. Balancing error here would be a generalization of the situation that occurs in poststratification when an overcount on the P-sample side that would normally balance with an undercount on the E-sample side no longer balances due to the P- and E-sample cases being included in different poststrata as a result of the use of different stratification variables.

This situation occurred with the 2000 census coverage measurement program, when the poststratification used in A.C.E. revision II was modified from that used in the original A.C.E. The Census Bureau decided to include two new factors in the poststratification that were available only for the E-sample due to their predictive power, resulting in different poststrata for the E-sample and the P-sample, believing that the new factors would provide preferable poststrata for estimation of the probability of correct enumeration status. The new factors were a variable indicating whether the E-sample enumeration was or was not a proxy enumeration and a variable indicating the type of census return (early mail return, late mail return, early nonmail return, and late nonmail return). While the addition of these variables to the E-sample poststrata certainly improved the partitioning of the E-sample into more homogeneous groups to reduce correlation bias, there was a concern that the difference in poststrata for the E-sample and the P-sample might cause a substantial number of failures in the balancing assumption. For example, a proxy interview often results in an enumeration with insufficient information for matching. Insufficient information enumerations were treated in 2000 as if they were erroneous enumerations, and the P-sample enumerations that would have matched to those cases were treated as census omissions. Since the E-sample cases were proxy enumerations and therefore placed in a poststratum that did not exist for the corresponding P-sample cases for A.C.E. revision II, these errors would be unlikely to balance.

A related issue can arise in the application of logistic regression models of both the match rate and the correct enumeration rate, but it is substantially more difficult to assess. If the variables differ for these two logistic regression models, coverage rate estimates for some combinations of these variables might be biased, although it is not known whether this would cause bias for the domains (defined by geography, age, race/ethnicity, etc.) that are of interest for census estimation. It is therefore important to determine when the benefit of improved predictive power for one of the two logistic regression models would outweigh the loss from the lack of balance.

Differences in the covariates used for the two logistic regression models could arise for two reasons. First, as mentioned above, some variables are not available for both the E-sample and the P-sample, since the P-sample interview is much more detailed. Second, even when the same variables are available for both the P-sample and the E-sample, their degree of predictiveness could differ markedly between the two logistic regression models. In the second case, by insisting on the use of identical predictors one may be sacrificing additional predictive power. Given that, if the balancing error is discovered to be relatively modest, one might still have a good reason to use different predictors in the two logistic regression models.

As an example, suppose erroneous enumerations were much more frequent for those enumerated using personal follow-up interviews in comparison with those enumerated using mailed back questionnaires (which was *not* the case in 2000). If there is no reason to believe that the same distinction would hold for omissions, it would then be sensible to include a variable that measured the rate of mail return as a covariate for the logistic regression model of correct enumeration status, but not to include it as a covariate in the logistic regression model for match status.

This problem may be reduced in the 2010 census given the collection of more data on census residence, the removal of duplicates during the census, other data improvement processes, and given the improved matching of KE cases (for the definition see the section below on missing data). To assess this tradeoff, one has to evaluate the quality of the resulting dual-systems estimates; to do that, the Census Bureau may have to make use of artificial populations analysis in which the true counts are known.

One way of satisfying this restriction for having the same predictors for both logistic regression models—but still retaining some of the predictive power of the excluded predictors—is to use predictors that are tabulated at the area level rather than at the individual level. This would therefore permit the use of predictors that were process or composition measures for small areas, which might provide substantial additional predictive performance. Finally, it should be stressed that this balancing problem is only relevant to the estimation of net coverage error—it does not arise in modeling the frequency of components of census coverage error.

The panel supports further work in developing logistic regression models, given their promise, particularly in looking for the benefits of additional covariates (again, including transformations and interactions). We add two additional possibilities for broadening the approaches under consideration that the Census Bureau may wish to consider. First, it is not necessary that one logistic regression model be used nationwide. Different regression coefficients and even different predictors could be used for different geographic or demographic domains. Second, logistic regression is only one of many statistical models that predict a dichotomous dependent variable. This is a discriminant analysis problem, and there are a number of more flexible methods—such as classification and regression trees, recursive partitioning, support vector machines, and modeling with flexible link functions—that have been shown to have applicability in practice. For instance, classification trees develop a tree structure of decision rules that select increasingly refined subsets of a population of interest, for which the subsets are identified by the joint range of values defined by the possible predictors and at each stage the subsets are selected to best discriminate between those that match or are correct enumerations and the remainder. Such an approach avoids the assumption of linearity used in logistic regression modeling and is therefore more flexible. Recent work by Wang et al. (2006) supports the benefits of such an approach in forming poststrata. Even if such an approach was not used in a production capacity in 2010, new information about what types of people or addresses fail to match in the census might be discovered through use of these techniques.

In 1999 the Census Bureau examined the ability of classification trees to create poststrata in comparison with the variables ultimately used in 2000. The poststrata produced did not perform any better than the ones used. Classification trees were also used to determine adjustment cells for unresolved cases using data from the 1998 dress rehearsal. In this case, the use of an additional variable, people aged 18–29 living at home, was discovered to provide some benefit. It is common to find that the benefits in going from logistic regression to these more flexible approaches are modest. However, the work the Census Bureau has carried out to date, though closely related, is not identical to what is being suggested. Furthermore, by examining alternative approaches, the Census Bureau will learn more about the patterns in the data at a very small cost since no additional data collection would be needed.

In addition, consistent with the new priority for coverage measurement in 2010, it is important that the Census Bureau consider directing more resources now to another group of discriminant analysis problems in addition to modeling match rate and correct enumeration rate. The search for correlates of components of census coverage error—the first step in developing a feedback loop for census improvement—will involve modeling the frequency of census omissions, erroneous enumerations, duplications, and enumerations in the wrong location (for various definitions of wrong location). Again, these are efforts to discriminate between membership in two groups using a set of predictors and can therefore be addressed through use of logistic regression or the other approaches listed above.

In such a research effort, the predictors that are clearly effective in the logistic regression models for match status and correction enumeration status may not be the same predictors that are effective in modeling the components of census coverage error. We describe the various types of predictors that should be considered for use in these models in Chapter 5. The key point is that the Census Bureau needs to devote considerable effort to determining both the best general form of statistical model for the four components of coverage error (i.e., logistic regression, classification tree, etc.) and which predictors are most effective in modeling the different dependent variables. The predictors that show predictive power are then potential candidates for causal factors of census coverage errors.

MISSING DATA IN NET COVERAGE ERROR MODELS

Missing data touches many aspects of DSE. In this section, we present principles for handling missing data and discuss those principles in the context of one problem, logistic regression modeling of the match rate.

Missing data affects net coverage error models in the following areas:

- P-sample noninterviews: households in the P-sample that either could not be found at home to interview or were unwilling to cooperate;
- missing characteristics in P-sample interviews: households for which the answers to some questions were not provided;
- unresolved P-sample match status: the in-person follow-up interview was unsuccessful in determining whether or not there was an E-sample match to a P-sample individual, often due to an incomplete interview;
- unresolved P-sample residence status: the in-person follow-up interview was unsuccessful in determining where that person should have been enumerated, often due to an incomplete interview;
- unresolved E-sample enumeration status: the in-person follow-up interview was unsuccessful in determining whether a person should or should not have been enumerated in the census, often due to an incomplete interview;
- missing data for individuals not in the P-sample: missing characteristics information used as covariates in the logistic regression models; and
- missing data for component errors: assessment of duplicate status, erroneous enumeration status, or whether someone was counted in the wrong place can be missing, often due to incomplete interviews.

These missing data problems are currently addressed by some form of imputation. Four general principles of imputation are worth bearing in mind when assessing and refining current approaches (see Little and Rubin, 2002:Chapters 4–5). First, imputations of sets of missing variables should be *multivariate*, to preserve relationships between them. Second, one should use *draws* rather than means: imputing draws from a predictive distribution of missing values rather than means avoids bias for estimates of nonlinear quantities (as in CCM equations); the hot-deck imputation method is a form of a draw. Third, imputations should be *conditional on predictive covariates* to the extent possible on all predictive auxiliary information. Fourth, standard errors should *incorporate measures of imputation uncertainty*, using multiple imputation or replication methods.

From the perspective of these principles, current imputations based on the hot deck are draws, and there is some attempt to condition on covariate information, although the full range of such information is not always exploited (see below). It does not appear that imputation uncertainty is included in estimates of standard errors, and the question of whether imputations preserve associations needs some attention.

To illustrate the application of these principles, we consider the problem of missing data in the context of estimation of the match rate, since the assessment of match status plays a fundamental role in DSE. We believe that many of the ideas are also applicable to the other missing data problems mentioned above.

Currently, the Census Bureau first imputes missing characteristics for the P-sample interviews. Next, using those imputed values along with the collected P-sample values and a before-follow-up match code, a logistic regression model is used to impute match status. As in most situations involving the treatment of missing data, the properties of proposed missing data treatments need to be assessed in the context of the complete data problem. Here, the complete data problem is the matching of the P-file and the E-file, when the P-file consists of the data collected for the individuals represented in the P-sample, and the E-file consists of the data collected for the individuals represented in the E-sample, namely those people counted in the census at residences in the PES block clusters. Figure 4-1 depicts the matching problem. In this figure, X represents variables used in the logistic regression (e.g., age, sex, race, and ownership status), Z represents auxiliary variables that are not in the logistic regression models but could be used in the matching operation (e.g.,

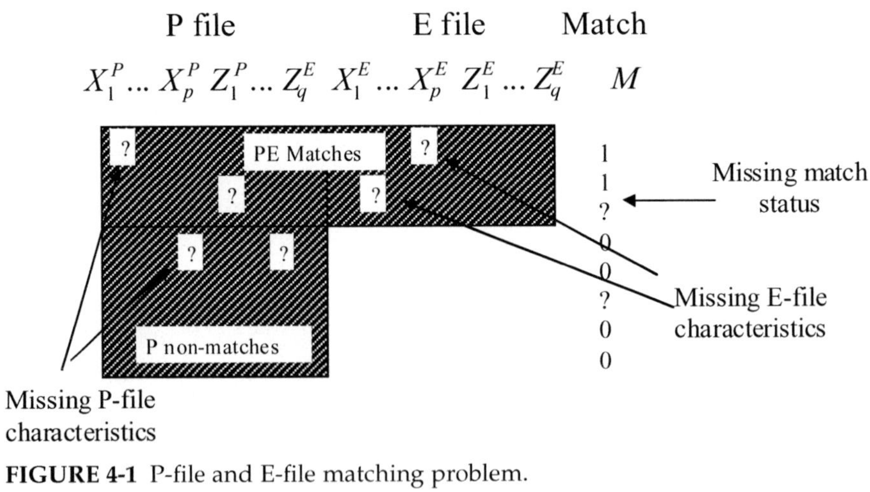

FIGURE 4-1 P-file and E-file matching problem.

name and detailed date of birth), and match status, M, is a binary function $M\left(X_i^P, Z_i^P, X_i^E, Z_i^E\right)$ of the characteristics for individuals in the E- and P-files (using an obvious notation).

The current procedure uses logistic regression with M as the dependent variable and X^P as the set of potential independent variables, augmented by the before-follow-up match code, which makes some use of X^E. Two complications that are ignored for the purposes of this discussion but need to be incorporated in a more complete analysis are the modeling of the probability of incorrect enumerations, which are modeled as another binary variable, and the fact that there is clustering of match status within households. (Cowan and Malec, 1986, provide one method for addressing the latter issue.)

Missing information occurs in two respects. First, some of the *X*s and *Z*s are missing for both the E-file and the P-file records for various reasons. These missing data complicate the problem of assessing match status. Second, partly due to the missing *X*s and *Z*s, the derived variable M is sometimes missing. The Census Bureau tends to "compartmentalize" these missing data problems—first solving the problem of missing *X*s and *Z*s, and then addressing the problem of missing *M*s—in estimating the parameters for the logistic regression model for match status.

However, it is better to conceptualize the problem as multivariate missing data, since fully effective imputation for missing data needs to preserve the relationships between various sets of missing and non-missing values for all the variables that have missing data. It also helps to specify the complete set of potential auxiliary variables and to elucidate the multivariate missing data patterns involving the dependent, independent, and auxiliary variables because it is important to preserve associations between variables that are frequently both missing, and missing *M*s are strongly related to missing *X*s and *Z*s.

The goal here is to improve estimates of the regression coefficients of the logistic regression of M on X by the effective use of cases with incomplete data. The key issue to determine is the amount of information in the incomplete cases. If there is effectively no useful information, these cases should be dropped, since imputation does not improve the estimates. We consider separately those cases with M observed and those cases with M missing.

If M is observed, cases with components of X missing have the potential to contribute substantial information to the regression. By the conditionality principle (essentially, any inference should depend only on the observed outcomes and not on any unobserved outcomes that might have been observed), imputations of components of X should condition on match status, which is important to avoid bias, and on any auxiliary variables that would help to predict missing *X*s.

If, however, M is missing, the cases will contribute little to the quality of the logistic regression coefficients unless M is imputed using auxiliary variables Z that are strongly predictive of M. If such information is available, it is important to condition on it in the imputation model. If not, imputing cases with M missing simply adds noise. Auxiliary information could, in principle, include data from the E-sample (in particular, data from the closest potential match). The Census Bureau's use of the before-follow-up match code is a good example of the implementation of this principle.

For the problem of imputation of match status, M, consider the following three possibilities for sets of conditioning information for imputing match status:

1. condition on $X^P_{obs,i}$: the observed predictors from the P-file in the logistic regression model for match status. This corresponds to the current procedure.
2. condition on $X^P_{obs,i}$, $Z^P_{obs,i}$: the observed predictors in the logistic regression model for match status, and the observed auxiliary variables used for matching
3. condition on $X^P_{obs,i}$, $Z^P_{obs,i}$, and $\left\{\left(X^E_{obs,j}, Z^E_{obs,j}\right)\right\} j \in S_i$: the set of best matches, which adds the information in the E-file of the best matches to case i.

These three alternatives, which differ with respect to the information on which to base an imputation of match status, can yield very different imputations. Conditioning on more variables is preferred if the additional variables improve the prediction of M.

To demonstrate the advantages of alternative 2 with the intermediate degree of conditioning, one can imagine a number of situations involving name, which is typically used to determine whether records match but is not typically suggested for use in the logistic regression model to impute match status. Situations range from those in which the E-file and P-file records have the same name, some with spelling inconsistencies but the names appear to match, and some without a name on one or both the E-file, the P-file, or both. It is likely that the logistic regression-based probabilities of a match should be different for these various situations. Furthermore, one can exploit any unused information to check on one's assumptions.

To demonstrate the advantages of alternative 3, which is the most comprehensive form of conditioning in which all relevant variables are used, consider the following example of a P-sample case with race missing. Also assume that this case matches to an E-sample case that has race identified as "Asian or Pacific Islander." An imputation that does not condition on E-sample information may incorrectly impute a race other than Asian (likely white). Even a close *potential match* to an E-sample case with Asian or Pacific Islander as the race increases the probability that the P-sample case has that racial characteristic. As another example, suppose one has a record with a name from the P-sample, and 10 records with very similar characteristics but without names from the E-sample (actually, the census records in the P-sample blocks). It is likely that one of those 10 records matches the P-sample record—or certainly more so than in situations in which no E-sample records have similar characteristics.

The Census Bureau appears to be assessing match status, in which there is missing information for various characteristics, based on both P-file and E-file information: however, the Bureau's logistic regression model for the imputation of match status uses P-file information and a limited amount of E-file information through use of the before-follow-up match codes. This approach could result in too many imputed non-matches. The distinction between the E-file and the P-file is unnecessary: Both files are datasets relative to the same block cluster, so they cover groups of people that overlap substantially—i.e., they are characteristics for primarily the same set of people. Although treating the imputation of match status and person characteristics as two separate problems is simpler, it violates the principle that imputations should be multivariate and appears to make inferior use of the information available. If there is some computational advantage to separating the problems in a sequential approach, one can still obtain the benefits of a full multivariate approach by conditioning as one proceeds from the first problem to the second.

A crucial assumption underlying the above is that the missing data are missing at random. Let R represent the missing data pattern that is observed, and let $V = \{X, Z\}$ be the variables that are conditioned on when using an imputation method. Then the assumption is specifically that:

$$\Pr\left(M, X^P_{mis} \middle| V, R\right) = \Pr\left(M, X^P_{mis} \middle| V\right) \text{ (M missing), and}$$

$$\Pr\left(X^P_{mis} \middle| M, V, R\right) = \Pr\left(X^P_{mis} \middle| V, M\right) \text{ (M observed and conditioned).}$$

Conceptually, the first equation depicts that, in imputing for logistic regression modeling, which entails representing the relationship between M and X^P_{mis}, conditioning on the missing data pattern adds nothing when already conditioning on V. The second equation depicts that in imputing for missing covariates, conditioning on the missing data pattern adds nothing when already conditioning on M and V. These formulas illustrate one of the underlying principles, which is that by conditioning on V, the assumption of missing at random is weaker than the assumption without this conditioning. (The stronger assumption is missing completely at random, which is that the missing values have the same distribution as the observed values, without any conditioning. This assumption is extremely unlikely to obtain in either missing data problem.) If the missing at random assumption is not considered reasonable, one could impute *M* by conditioning on various aspects of *R* (referred to as pattern-mixture models; for details, see Little, 1993).

It would be valuable for the Census Bureau to assess its current imputation methods in its coverage measurement models for consistency with the above principles. As noted above, the logistic regression approach for modeling match status seems too focused on the P-file data, ignoring potentially useful information both in auxiliary data used in the matching algorithm and in the E-file. It may be that after this reconsideration, modest adjustments to the current procedures will provide a model for match status with smaller mean-squared error under a variety of realistic models for both the generation of data and missing values.

The Census Bureau's current imputation methodology, hot-deck imputation, works well in situations with limited covariate information. However, the difficulty in this approach is that dealing with more than a few covariates at a time compromises its ability to condition on all relevant variables. In contrast, parametric multiple imputation methods make better use of covariate information, and these methods can be used to estimate the contribution to variance as a result of the missing data. An example of this is IVEWARE (see, e.g., Raghunathan et al., 2002). Another question involves the role of imputation for missing census characteristics values. After estimating the logistic regression of *M* on *X*, imputations for missing census characteristics are needed to provide the predictors for input to the logistic regression models to estimate a match probability for these cases, through:

$$\Pr\left(M \middle| X_{obs}^{E}\right) \approx \Pr\left(M \middle| X_{obs}^{E}, \hat{X}_{mis}^{E}\right)$$

as input into the small-domains estimation procedure to be used in 2010. Use of hot-deck imputation here is reasonable, but an alternative is to estimate this probability directly given the observed E-sample characteristics,

$$\Pr\left(M \middle| X_{obs}^{E}\right).$$

This approach avoids the additional uncertainty from the imputation, and it should be straightforward to employ with the move to use of logistic regression.

Finally we note that the coverage measurement data collected in 2010, in particular the various follow-up data collections that are typically carried out, could be used to validate the imputation models used, though the sparseness of these samples may make this of only limited utility.

In summary, missing data methodology needs to be viewed in the context of the complete-data problem, file matching. As noted above:

- Imputation is only useful if it adds information to the logistic regression; otherwise cases can be dropped.
- Imputations should be multivariate in order to preserve associations between missing variables.
- Imputations should condition on predictive covariates.

For example, imputations should condition on M if M is observed, and imputations should condition on potential covariate information from matches or potential matches from the E-file. Some form of weighting might be developed to reflect the strength of the potential matches. The Census Bureau could also consider parametric multiple imputation as an alternative to the hot deck because it makes better use of the covariate information and because it propagates imputation uncertainty. Finally, the Census Bureau could also consider nonignorable models, such as pattern-mixture models, if the missing-at-random assumption is likely to be violated.

This is a set of research problems that the Census Bureau needs to allocate substantial staff resources to address. We believe that the benefits are likely to be considerable and the understanding from the P-sample matching problem discussed in detail should be transferable to some of the other missing data problems listed on p. 103. The Census Bureau should identify missing data methods that are consistent with the philosophy that is articulated above and implement those methods in support of statistical models of Census Coverage Measurement data in 2010.

Recommendation 7: The Census Bureau should develop missing data techniques, in collaboration with external experts if needed, that preserve associations between imputed and observed variables, condition on variables that are predictive of the missing values, and incorporate imputation uncertainty into estimates of standard errors. These ideas should be utilized in modeling the census coverage measurement data collected in the 2010 census.

MATCHING CASES WITH MINIMAL INFORMATION

For an E-sample enumeration to have sufficient information for matching and follow-up, as defined in the 2000 census, it needed to include a person's complete name and two other nonimputed characteristics. To be data defined in the census itself, an enumeration simply had to have two non-imputed characteristics. In the A.C.E. E-sample in 2000, 1.7 percent

(4.8 million sample survey weighted) of the data-defined enumerations had insufficient information for matching and follow-up. These cases were coded as "KE" cases in A.C.E. processing.

A.C.E. estimation treated KEs as having insufficient information for matching, and they were removed from the census enumerations prior to dual-systems computations. If KEs are similar in all important respects to census enumerations with sufficient information for matching, removal from dual-systems computations slightly increases the variance of the resulting estimates, but it does not greatly affect the estimates themselves. Removal of KEs helped to avoid counting people twice because matches for these cases are difficult to ascertain. Also, it was difficult to follow up these E-sample cases to determine their match status if they were initially not matched to the P-sample because of the lack of information about whom to interview.

However, some unknown and possibly a large fraction of these cases were correct enumerations. Therefore, removing these cases from the matching inflated the estimate of erroneous enumerations, and it also inflated the estimate of the number of census omissions by about the same amount, since roughly the same number that are correct enumerations would have matched to P-sample enumerations. (There is no way of validating this assumption since the KEs generally cannot be followed up.) Given that the emphasis in 2000 was on the estimation of net census error, this inflation of the estimates of the rates of erroneous enumeration and omission was of only minor concern. However, with the new focus in 2010 on estimates of components of census coverage error, there is a greater need to find alternative methods for treating KE enumerations. One possibility that the Census Bureau has explored is whether many of these cases can be matched to the P-sample data using information from other household members.

To examine this possibility, the Census Bureau carried out an analysis using 2000 census data on 13,360 unweighted data-defined census records with insufficient information for matching to determine whether their match status could be reliably determined. (For details, see Auer, 2004; Shoemaker, 2005.) This clerical operation used name, date of birth, household composition, address, and other characteristics to match the cases to the P-sample. For the 2000 A.C.E. data, 44 percent of the KE cases examined were determined to match to a person who lived at the same address on Census Day and was not otherwise counted, with either "high confidence" or "medium confidence" (which were reasonable and objectively defined categories of credibility). For the 2000 census, this would have reclassified more than 2 million census enumerations from erroneous to correct enumerations, as well as a similar number from P-sample omissions to matches, thereby greatly reducing the estimated number of census component coverage errors.[11] (We note that it is important in carrying out this research to remain evenhanded in evaluating whether a case does or does match; this is not simply an effort to identify more cases that are matches.)

The treatment of the KEs remaining after this revisiting of the definition of insufficient information for matching can be viewed as another component of "error" in the same way that a person incorrectly geocoded is an error—that is, as a problem for processing but not a part of what one would call an omission or an erroneous enumeration. Therefore, the use of the term "erroneous enumeration" for these cases is inappropriate. Cases with insufficient information should be treated as having unknown or uncertain enumeration or match status and the term "erroneous" should be reserved for incorrect enumerations. The terminology used needs to distinguish between types of error and the uncertainty associated with these types of error for particular cases.

The panel is impressed with the findings of this research, which should substantially improve the assessment of components of census coverage error in 2010. In considering further development of the idea, it would be useful to try to find out more about any characteristics associated with KEs in order to find out how to reduce their occurrence in the first place. StARS might be useful for this purpose. Furthermore, the clerical operation used to determine the status of KEs was resource intensive, and it would be useful to try to automate some of the matching to reduce the size of this clerical operation in 2010.

We anticipate that, as a result of this research, the Census Bureau will adopt a different standard of what is considered to be insufficient information for matching more generally.

DEMOGRAPHIC ANALYSIS

Demographic analysis may be facing a very dynamic period in the next few years for several reasons. First, nearly all record systems are becoming increasingly more complete with higher quality data. Second, the American Community Survey is now providing a great deal of useful, subnational information that could be used to improve and extend demographic analysis estimates. Third, StARS, a merged, unduplicated list of U.S. residents and addresses, is also a likely source of information on the number of housing units and residents at small levels of geographic aggregation that could also be used to improve demographic analysis estimates.

---

[11]For the remaining unresolved cases, the Census Bureau currently plans to treat them in a separate category as "enumerations unable to evaluate."

At the same time, some things are becoming more complicated, notably, the expansion of the number of race and ethnicity categories on the decennial census and the growing and increasingly mobile population of undocumented immigrants.

In this context, the panel was asked to examine how demographic analysis might function more effectively as an independent assessment of the quality of the coverage of the decennial census. In addition, the panel was asked to consider the use of sex ratios from demographic analysis, especially for Hispanic residents, to reduce the effect of correlation bias in dual-systems estimation.

As described above, the basic demographic analysis equation is

$$\hat{P}_{ij}^{NEW} = \hat{P}_{ij}^{OLD} + B_{ij} - D_{ij} + I_{ij} - E_{ij},$$

where $\hat{P}_{ij}^{NEW}$ represents the current estimate of the population for demographic group i and geographic area j, $\hat{P}_{ij}^{OLD}$ is the analogous estimate for a previous census, B_{ij} represents the number of births between the current and a previous census, D_{ij} represents the number of deaths between the current and a previous census, I_{ij} represents the number of immigrants between the current and a previous census, and E_{ij} represents the number of emigrants between the current and a previous census, all for demographic group i and geographic area j. Once $\hat{P}_{ij}^{NEW}$ is computed, the net census undercount, U_{ij} for demographic group i and area j is defined as $\hat{U}_{ij} = \hat{P}_{ij}^{NEW} - C_{ij}$, where C_{ij} is the census count, again for demographic group i and area j.

Error is introduced into estimates from demographic analysis due to omissions in the birth and death records and due to large inaccuracies in the data on immigration and emigration. The error in net undercoverage estimates from demographic analysis then stems from error in the various components, error in the census counts, and any lack of alignment of the demographic categories.

Given these concerns, the most reliable outputs from demographic analysis are any national counts by age and sex, and functions of such counts, in particular sex ratios by age; birth and death estimates; and historical patterns of various kinds. More problematic outputs are race (depending on the degree of alignment to the new race/ethnicity categories) and subnational estimates for demographic groups.

The most problematic outputs are estimates of international migration components, estimates of the Hispanic population, subnational totals for states and smaller geographic areas.

The Census Bureau plans for demographic analysis in 2010 are to produce "estimates" and "benchmarks," with estimates represented to users as being more reliable than benchmarks. The Census Bureau will produce estimates of national level totals by year of age and by sex, and estimates of 2000–2010 change for the above groups. The Census Bureau will also produce benchmarks of national net undercount error by age, sex, and race/ethnicity. In addition, the demographic analysis program will produce sex ratios by age and race/ethnic origin, possibly for use in reducing the effects of correlation bias on estimates of net undercoverage from the census coverage measurement program.

Even without any major advances from 2000, demographic analysis will still likely play an important role in evaluation of the 2010 census. As pointed out above, demographic analysis provided an early indication that the initial estimates of the total U.S. population from A.C.E. may have been too high, and it will continue to provide an estimated count that serves as a useful estimate for many demographic groups and a useful lower bound for others.

The Census Bureau is currently pursuing important research directions, though it is unclear whether they will contribute to the 2010 demographic analysis program. Those research plans include: (1) improved estimation of international migration, (2) estimation of the uncertainty of demographic analysis estimates, and (3) progress towards the production of subnational estimates. The latter includes research on methods and data sources, with some pieces already considered of possibly acceptable quality, such as estimates of the number of people younger than 10 years of age at the state level. We believe that these are extremely important projects to pursue and deserve full support from the Census Bureau.

In addition, the panel has the following questions concerning the 2010 demographic analysis estimates that may help orient these research avenues:

- Given that there is race/ethnicity incomparability between the decennial census and demographic analysis, which categories are going to be used in 2010?
- Given overlapping data for some cohorts (e.g., Medicare information for those over 65) in comparison with standard demographic analysis, which sources will be used and how will that be determined? Will there be efforts to combine information?
- Estimates of Hispanic origin were produced by the censuses of 1980, 1990, and 2000, as were adjusted counts. Have these sequences been examined to determine their likely quality over time?
- In considering subnational estimates, relatively high-quality estimates are available of the number of native-born children under 10 years old at the state level, and the number over 65 from Medicare, again at the state level. Given additional information on interstate migration from tax returns, school enrollment, and possibly the American Community Survey, could high-quality

estimates be provided for the remaining demographic groups at the state level?

- If the Census Bureau again uses sex ratios from demographic analysis to reduce the correlation bias in adjusted population counts, should these be applied for all minority men or selectively, as in 2000?
- The American Community Survey is providing information that might be extremely useful for improving demographic analysis estimates. The possibilities include: (1) better estimates of the number of foreign-born residents, (2) better estimation of net international migration, and (3) information on sex ratios for more detailed ethnic and racial groups. How should each of these information sources be best used to improve demographic analysis, and what evaluations should be used to support decisions of implementation?
- Measurement of the size of the undocumented population is a continuing problem for demographic analysis. The current method, described in Passel (2005), is, roughly speaking, to subtract the estimated size of the legal immigrant population from the estimated size of the total foreign-born population. Are there any new methods that might be more effective in estimating the size of this population?
- StARS is already, or will soon be, of high enough quality to provide useful input into demographic analysis estimates. There are reasons to believe that administrative records could play an important role in improving various aspects of demographic analysis, especially the counting of immigration and emigration, and research in this area would be very desirable.

Demographic Analysis in Combination with Dual-Systems Estimation

As part of A.C.E. revision II, the Census Bureau decided to modify the final A.C.E. estimates based on sex ratios from demographic analysis and the assumption that the A.C.E. counts for women and children were correct. Specifically, at the level of aggregate poststrata (aggregated over nondemographic and geographic characteristics), the A.C.E. counts for black men 18 and over and for all other males 30 and over were adjusted upward so that the ratio of women to men for A.C.E. (essentially) agreed with that estimated using demographic analysis.

The argument in support of this joint use of demographic analysis and dual-systems estimation is as follows. Demographers generally believe that the most accurate outputs of demographic analysis are national-level sex ratios by age for blacks and nonblacks. Even if absolute counts are subject to some bias, sex ratios are expected to be quite accurate.

Historically, at least for adult blacks, the corresponding male-to-female ratios based on adjusted counts using dual-systems estimation have been lower than those from demographic analysis, suggesting that correlation bias (or other sources of bias) result in relative underestimation of adult males by dual-systems estimation. Because the most obvious source of correlation bias (heterogeneity of enumeration probabilities) would not have resulted in a negative bias for dual systems estimates, the most conservative step, in terms of additional counts, is to leave estimates for the female population unchanged and to increase the male population enough so that the resulting sex ratios for the adjusted counts agree with those from demographic analysis.

It is not sufficient to simply add these additional enumerations at the level of the aggregate poststrata; they must then be allocated down to the poststrata within each of the aggregate poststrata. Bell (1993) and Bell et al. (1996) identified five different methods for doing so, but there is little evidence available as to which of the methods works best. The Census Bureau selected one of these five approaches on the basis of its best judgment, but the arbitrariness of the selection, along with the fact that the counts were sensitive to the method used, is troubling. Also, given the limitations of demographic analysis, this technique could not be applied to such particular subgroups as nonblack men aged 30 and over (especially Hispanics), despite some historical evidence that a similar correction might have improved estimates for those subgroups. Finally, adjusted counts for both adult males and females have rested on the assumption that there is no correlation bias for adult females.

Admittedly, the approach used resulted in higher "face validity" for the adjusted census counts at the aggregate level as a result of the consistency with the sex ratios from demographic analysis. However, given the issues described above, especially the lack of a formal assessment of the effect of this process on the quality of the resulting counts, the decision was controversial.

Given this situation, it seems reasonable to carry out a more comprehensive evaluation of what was done in 2000 and possible alternatives before adopting a similar modification in 2010. (The Census Bureau currently plans to use a similar technique in 2010 as a correction for correlation bias.) Artificial population studies, in which models are developed to designate which individuals in an artificial population are and are not missed by the census, the PES, and by the record systems used by demographic analysis could be useful in such evaluations. We suggest that the Census Bureau include the approach described by Elliott and Little (2000) in their analysis of the method used in 2000. Their approach provides useful smoothing to the technique described in Bell (1993) and Bell et al. (1996). In addition to the beneficial smoothing, Elliott and

Little's work provides estimates of precision that incorporate the uncertainty in the demographic analysis sex ratios.

In addition, the information from the American Community Survey and from StARS on various demographic statistics, such as sex ratios, could be considered for use in providing not only modifications to the counts for males, but also modifications to the counts for females, avoiding the necessity of relying on the assumption that no correlation bias exists for that demographic group.

Estimation of Uncertainty of Demographic Analysis

The Census Bureau (see Robinson et al., 1993) conducted initial research on developing uncertainty intervals for population forecasts, but to date these have not been fully developed. Development of such uncertainty intervals would have two benefits: users would be supplied with uncertainty intervals with a formal probabilistic interpretation, and estimates from demographic analysis could be combined with estimates from independent sources by weighting by the precision of each estimate.

In the past 15 years, a number of researchers have suggested interesting methods to consider for development of uncertainty intervals. Poole and Raftery (2000) suggest the use of Bayesian melding for this purpose. Briefly, the idea is that one has expert knowledge about inputs to a deterministic model and their variability (i.e., a prior distribution) and expert knowledge about the outputs of interest (the forecasts), which through exact or approximate inversion presents a second prior distribution for the inputs. These two prior distributions then have to be reconciled. There is also the most recent data for the inputs that have been collected, and one can develop likelihoods for the previous inputs and outputs given the data. Bayes rule is then used, implemented by the sampling-importance-resampling algorithm of Rubin (1988), to update the prior distribution to produce a posterior distribution of the forecasts, which would include a posterior variance.

Other approaches have also been suggested by, among others, Alho and Spencer (1997) and Lee and Tuljapurkar (1994). Given all of this promising research, and the benefits from the development of uncertainty intervals, it would be valuable for the Census Bureau to revisit this issue and evaluate some of these approaches for their applicability to demographic analysis of the U.S. census. It is true that the U.S. census tends to have idiosyncratic challenges each decade, such as the number of undocumented immigrants that are enumerated in a given census, the number of duplicate enumerations from multiple modes of enumeration, or the degree of census undercoverage, and these challenges may be difficult to model. Therefore, in particular, the specific stochastic models suggested by Alho and Spencer (1997) and Lee and Tuljapurkar (1994) might need some modification. However, even recognizing this, if started now, the panel is confident that a research effort devoted to this issue would very likely produce useful uncertainty intervals for the 2010 census.

In summary, demographic analysis played an important role in helping to evaluate the estimates produced by A.C.E. in 2000, and it can play an even larger role in 2010 and 2020, especially if some improvements are implemented. Those improvements include improving the measurement of undocumented and documented immigration, development of subnational geographic estimates, development of estimates of uncertainty, and further refining methods for combining demographic analysis and coverage measurement survey information.

Recommendation 8: The Census Bureau should give priority to research on improving demographic analysis in the four areas: (1) improving the measurement of undocumented and documented immigrants, (2) development of subnational geographic estimates, (3) assessment of the uncertainty of estimates from demographic analysis, and (4) refining methods for combining estimates from demographic analysis and postenumeration survey data.

5

Analytic Use of Coverage Measurement Data

After the 1990 and 2000 censuses, coverage measurement tabulations focused on estimates of net coverage error for poststrata and aggregates of poststrata. While such tabulations remain valuable, the expanded goals for census coverage measurement (CCM) in 2010 call for a wider array of tabulations and analyses. First, there should be tabulations that break out the four types of census errors (defined in Chapter 2). Second, in order to learn about how census operations affect the occurrence of census errors, each type of error should be linked with a much wider array of variables that include detailed measures of census processes. Finally, the Census Bureau needs to perform analyses that take advantage of the wealth of information from CCM. This chapter first presents a framework for defining, estimating, and modeling the components of census coverage error. It then broadly specifies the variables that might play a role in statistical models. Lastly, it considers data products in the light of the goal of census improvement.

FRAMEWORK FOR UNDERSTANDING COVERAGE ERRORS

We first consider a framework for defining and estimating components of census coverage error. An excellent start in this direction was provided in Mulry and Kostanich (2006), which details the assumptions underlying dual-systems estimation (DSE) in the presence of nonresponse and other census data deficiencies. In carrying this out, Mulry and Kostanich provide rigorous definitions for some of the components of census coverage error, which are the "dependent variables" for a research program modeling these components. (See Appendix A for some of the details of this research.) The panel supports this research and would like to see it developed more fully, both operationally and in broadening the structure to incorporate duplicates.

Recommendation 9: The Census Bureau should further develop and refine its framework for defining the four basic types of census coverage error and measuring their frequency of occurrence. The Census Bureau should also develop plans for operationalizing the measurement of these components using data from the census and the census coverage measurement program.

In addition to producing tabulations of net coverage error at some level roughly comparable to the poststrata used in 2000, the Census Bureau has started developing plans for producing tabulations of components of census coverage error, including the percentages of erroneous enumeration, duplicates, missed enumerations, and people counted in the wrong area (e.g., in the wrong state) by major demographic characteristics, by geography (possibly states), and by some census operations (such as mailout–mailback in comparison with alternatives). The panel believes that such tabulations, while representing a very positive shift in the role of products of coverage measurement, still fail to fully utilize the richness of information contained in the coverage measurement data that will be collected in 2010.

The hoped-for feedback loop linking component census coverage errors to their causes will require the creative use of exploratory data analysis to answer the following general question:

Which census processes are associated with a substantially increased rate of erroneous enumerations, duplications, omissions, or enumerations in the wrong location?

This, in turn, will necessitate answering several other general questions:

- *What types of housing units are missed more often than others?*
- *What types of housing units are duplicated more often than others?*
- *What types of people are missed more frequently than others?*
- *What types of people are erroneously enumerated more frequently than others?*
- *What types of people are duplicated more often than others?*
- *What types of people are counted in the wrong location more often than others?*

Through use of data analyses that use these outcome variables as dependent variables, one should be able to identify collective values for sets of predictors that identify situations that produce a higher rate of one (or more) of the four types of coverage error. In other words, given the current plans for 2010, it seems very possible that the Census Bureau could provide assessments of when a census enumeration is likely to be erroneous, when a census enumeration is likely to be a duplicate, when a census enumeration is likely to be counted in the wrong location, and when a P-sample enumeration is likely to be a census omission.

It is also reasonable to expect that beginning to develop answers to the above questions and provide the above assessments will be an important step toward identifying alternative census processes and designs that might reduce the frequency of the associated coverage errors in subsequent censuses. For example, missing people in gated communities might suggest altering the way nonresponse follow-up is carried out for such residences. Duplicating people serving in the military would suggest a change to the examples of residency examples provided on the census questionnaire.

Toward this end, we outline a general approach to the statistical modeling of component census coverage errors that we believe the Census Bureau should consider. Note that the statistical models outlined will be fit using data from the P-sample and from the associated E-sample enumerations, though efforts should be made to augment these models with data from administrative records data, the American Community Survey, and, if possible, other sources.

STATISTICAL MODELING

The following provides some specifics as to the variables that should be used in statistical models in support of census process improvement.

Dependent Variables

There are four primary types of dependent variables that should be used. The first dependent variable is an indicator variable as to whether a P-sample enumeration is a census omission. It may also be useful to consider separate indicator variables that indicate whether the omission is (a) a within-household omission, with others in the housing unit having been enumerated in the census; (b) a whole-household omission in a listed housing unit; or (c) an entirely missed housing unit.

The second dependent variable is an indicator variable as to whether an E-sample enumeration is a census duplicate. It may also be useful to consider separate indicator variables that denote whether the situation is (a) a whole-household duplicate or (b) a duplication of an individual in a household in which others were counted only once. Also, consider the possible ways in which people might be duplicated. People that are often duplicated are those in nursing homes, people in prisons, or people in military barracks being enumerated with their relatives rather than at their institutional residences. Also there are college students being enumerated where their parents reside and at school; people with winter homes being counted as residents at those locations and also at their primary residences; and movers being counted both at their previous residence and at their current residence. These various causes of duplication result in different displacements between the correct residence and the duplicate residence. Therefore, to help differentiate between these causes of duplication, it may be useful to have dependent variables that distinguish between duplications that have varying degrees of displacement between the two residences.

The third dependent variable is an indicator variable as to whether an E-sample enumeration is erroneous. It may also be useful to consider indicator variables that denote whether the erroneous enumeration corresponds to (a) a fictitious person, (b) someone who was born after Census Day, (c) someone who died before Census Day, or (d) a visitor.

The fourth dependent variable is an indicator variable as to whether an E-sample enumeration is in the wrong location. As mentioned above, there are different possible levels of geographic displacement (e.g., state or county), and for each definition a separate indicator variable could be formed. Further, it may also be useful to consider separate indicator variables that could distinguish between the following situations: (1) a household counted at the right address/housing unit but in the wrong place due to a geocoding error, (2) a household counted in the wrong housing unit, due to a move, and (3) a household counted at a Census Day residence that is not the usual residence.

Before continuing, we note that there will always be mistakes in attributing census component coverage error to one of the four types of component coverage error, and even whether a coverage error occurred at all. For example, due to matching errors, some cases judged to be census duplicates in the coverage measurement program through use of the national match for duplicates will not be duplicates and so a few correct enumerations will be judged as being duplicates. For the same reason, some actual duplicates will be erroneously classified as correct enumerations. Other situations of ambiguity and error will certainly exist in assessing erroneous enumerations, enumerations in the wrong place, and omissions. However, the current methods for judging which cases belong to which categories are sufficiently reliable to strongly support the current aims of the Census Bureau to improve census methodology.

Independent Variables

Primary

To understand which subset of individuals and housing units are more frequently subject to coverage errors of the four indicated types, and to understand what census processes contributed to those errors, it is necessary to focus on predictors that distinguish between individuals and housing units that are likely to have different interactions with census enumeration processes, as well as predictors that indicate the census processes that were used to attempt to enumerate those individuals and housing units.

We caution that the set of census component processes that one might want to represent is much richer than the discussion here would suggest. To get some sense of the complexity, there will be dozens of different questionnaires used in the 2010 census (to account for different forms of delivery, foreign languages, and other factors). Given the size of the postenumeration surveys (PES) used to date, the sample sizes of the coverage measurement program are unlikely to support analysis of the rarer component processes used in conjunction with the more detailed subsets of the population. Therefore, some compromise between full representation and parsimony in modeling must be struck.

We recognize that some of the covariates suggested for use in statistical modeling of the frequency of various components of coverage error may not be routinely available given the planned data collection in the P-sample blocks. For instance, we discuss below the possibility of determining whether someone in a household telephoned for questionnaire assistance and whether that was associated with a higher or lower rate of component coverage error. Such information has not been collected in previous coverage measurement programs and would therefore not be available to modelers. However, as mentioned above, if a master trace sample is collected in a way that overlaps substantially with the P-sample blocks in 2010, such covariates could be made available for analysis. Our outline assumes that certain covariates will be available in 2010. If this information will not be available, we hope that this discussion motivates a revision of plans to make more predictors available to support the models. With these caveats, we discuss the primary predictors that might be valuable to include in statistical models for component coverage errors.

Individuals and Housing Units There are four individual and housing units variables that need to be considered. First is the type of enumeration area and other features that identify the local geography and the types of housing units in the area. Housing units in the United States are primarily enumerated using one of three procedures: mailout–mailback, update–

leave, and list–enumerate (see National Research Council, 2000). Broadly speaking, housing units are selected for enumeration by one of these three procedures as a function of the quality of the mailing list and whether the housing units have unique identifiers. In addition, other features that might be related to coverage error are the frequency of small multi-unit residences in the area, the rates of new construction, and the rates of demolition of residences, as well as whether the immediate area is mainly residential or a mixture of residential and business establishments.

The second type of variable is the kind of housing unit. Indicator variables that might be useful to consider include whether the household in question has been newly constructed or is part of a small multi-unit building. Other indicator variables that are likely to be predictive are whether the housing unit is part of a group quarters and, if so, what kind of group quarters.

The third type of variable is people's demographic characteristics, which are related to differences in rates of components of coverage error (e.g., being between 18 and 22 is related to increases in the chances of duplication), and there also remains considerable interest in assessment of the quality of the census as a function of these characteristics. Therefore, it is important to include them in any statistical models of the rates of components of coverage error.

The fourth type of variable involves the relationships of residents. It is generally believed that a household with people who are unrelated to each other or are not part of the same nuclear family is related to census omissions. Therefore, indicator variables for this would likely be useful.

Census Processes At least five types of census process variables could be useful in these models. The first type uses results of the Master Address File (MAF) building process. Four basic types of coverage error can result from mistakes in the MAF building: if a nonresidential unit is erroneously included, if a housing unit is included twice, if a housing unit is omitted, and if an address is geocoded in the wrong location (which involves the interaction of the MAF and TIGER).[1] None of these mistakes forces a census coverage error, since these mistakes can be, and often are, corrected during the field enumeration. However, it seems clear that indicator variables of these errors are very likely to be associated with a greater frequency of coverage error. Therefore, four variables that should be considered for inclusion in statistical models to predict the relevant components of census coverage

---

[1]These variables, and others listed, may serve dual purposes. For some analyses they may be explanatory (or stratifying) variables. But it would also make sense to use them as outcomes in other analyses, modeled as functions of housing characteristics and MAF-related operations.

error are a variable that indicates when an address included in the MAF is not a residence, a variable that indicates when an address is included more than once on the MAF, a variable that indicates when an address was omitted from the MAF, and a variable that indicates when a geocoding error has been made. It is anticipated that these indicator variables will be strongly predictive of the obvious component coverage error. (It may sometimes be difficult to identify the specific address-building process that resulted in a MAF error: for example, knowing that an address was omitted from the MAF does not necessarily identify the specific address-building process that was in error.)

The second type of process variable to be considered uses results of the mailout/mailback enumeration. Three indicator variables related to the initial mail data collection might be associated with coverage errors: (1) an indicator that a foreign language questionnaire was requested or some other contact was made with telephone questionnaire assistance, (2) an indicator of some degree of item nonresponse on the returned questionnaire, and (3) an indicator that some of the responses to the census questionnaire needed to be keyed rather than scanned from image. Each of these indicator variables may be related to difficulties that respondents had in responding to the census questionnaire.

The third type of process variable to be considered uses information from nonresponse follow-up. Indicator variables associated with nonresponse follow-up that might be related to coverage error are the number of attempts needed to collect the information from initial mail nonrespondents and whether the ultimate enumeration was through a proxy respondent.

The fourth type of process variable uses information from the coverage follow-up interview. Indicator variables involving the coverage follow-up interview that might be related to coverage error are the six possible reasons for a coverage follow-up interview (more than six residents in the household, a count discrepancy in the census questionnaire, a possible duplicate enumeration given the national search for duplicates, count discrepancy with administrative records counts, response to the undercoverage probe, and response to the overcoverage probe). Another variable in this category is the number of phone calls needed to complete the coverage follow-up interview.

The fifth type of process variable involves other modes of enumeration. Other methods of being enumerated include "Be Counted" and through telephone questionnaire assistance. However, these and other modes are relatively rare, and it is doubtful that the PES would support statistical models that made use of variables that indicated their use. However, expansion of these modes of enumeration might take place, and in that case, indicator variables of their use might be useful in future models of the frequency of component coverage error. Similarly, any other process that occurs relatively frequently that is not included in this outline of census component processes, could be considered for inclusion.

Predictors Related to Enumerator Error There has been little research as to whether census coverage error is related to the effectiveness or doggedness of field enumerators. However, if possible, it would seem useful to examine. One variable that might discriminate between effective and ineffective enumerator effort is the turnover rate for the associated field office, and other variables that might be useful can be imagined, including variables that indicate how enumerators faired in any training exercises and variables that represent the results of the quality control program that checks on each enumerator's initial workload.

Contextual Factors

There are likely to be additional contextual variables that are associated with the frequency of components of census coverage error that may relate to how the census operates in a broad sense. Examples of such covariates are the percentage of people in an area that own their own residences, the local mail return rate, the local crime rate (assuming improvements are made to the completeness of the Uniform Crime Reporting system), the health of the local economy (using the local employment rate from the American Community Survey, for example), or the percentage of estimated undocumented local residents in the current or the previous census (again from the American Community Survey, though estimated indirectly).

Implementation

To facilitate use of these various dependent and predictor variables, it is important for the census coverage measurement database to be structured to enable the representation of data at multiple levels—including individual, household, local area, and the area covered by the Census' local enumeration office—since various variables will exist at these different levels. In addition, it will also be important to be able to link the E-sample data to the P-sample data for the same individual and other individuals in the same household.

It would be premature at this point to suggest the precise form of the statistical models that could be used for this application, but since most of the dependent variables are dichotomous, logistic regression, discriminant analysis, and classification trees are obvious models to consider. Moreover, it is very likely that given the complexity of the underlying phenomena being modeled, focusing on fixed-effects models with predictors restricted to the poststratification variables from 2000 will be unsatisfactory. Instead, what is needed is a representation of the complexity of the situation, involving characteristics of households, housing units, census processes, enumerators' performance, and interactions among these variables. In addition, the Census Bureau should also examine the possibility of using separate regression models for separate geographic domains.

The panel stresses that although the implicit orientation of this discussion has been on developing individual-level models, modeling at the individual level is not the only way of making progress in developing feedback loops for census improvement. For example, one could take the areas that have a high frequency of net undercoverage, or any of the four types of coverage error, and see if those areas have any similarities. In particular, one could carry out a cluster analysis of the areas with high rates of duplication to find out that, say, 30 percent of these areas are college towns, 35 percent are rural areas that have a lot of people with alternative addresses, with the remainder difficult to categorize. Such an analysis would provide obvious clues as to how to modify the census to reduce the duplication rate in the future. Or one could try to develop models or carry out exploratory data analysis at the household level.

There is a strong need for multiple approaches to the analysis of such data. It remains to be seen which kinds of exploratory data analyses or statistical modeling techniques are most helpful to the Census Bureau in identifying census deficiencies. It would be important to consider a wide variety of models and analyses before focusing on any small number of techniques. And it would be comforting if different approaches to the modeling produced similar findings. And, consistent with the need to maintain disclosure avoidance, it would be valuable to share the data with outside experts in various areas of application so that the latest techniques are examined for their applicability.

To summarize, in developing models for the frequency of the components of coverage error, and for modeling match rate and correct enumeration rate, the Census Bureau needs to consider a broad range of possible approaches before focusing on a specific model. Alternatives to the approach currently being considered include logistic regression with other link functions, discriminant analysis, and data mining, such as classification trees, support vector machines, and neural nets. Candidate regression models would include various subsets of predictors, as well as transformations and interaction terms. Also, separate models by geographic strata of various types should also be considered, for instance separate models for urban and rural areas.

We stress that the inclusion of interaction terms in such models is particularly important, since there are likely very significant interactions between geography and demography. We are unaware of any studies by the Census Bureau, in evaluating coverage error for previous censuses, where the possibility of complex regional effects has been investigated. For example, it is very easy to believe that the growing immigrant Spanish-speaking population will have a different response than others to changes in the presentation of the questions on race/ethnicity and residence. Also, changes to the procedures used to procure foreign language questionnaires and changes in the procedures used to hire field enumerators that are bilingual may also have an important impact on components of census coverage error.

The variables used for these various problems need not be identical, and therefore, separate model building research would need to be carried out. However, to avoid some form of balancing error, there may be a benefit to using the same variables for the logistic regression model of the match rate and erroneous enumeration rate in estimating net coverage error: the Census Bureau could investigate the costs and benefits of this idea prior to deciding whether to apply this constraint.

Recommendation 10: In developing the logistic regression models or other types of discriminant-analysis models of match status, correct enumeration status, and components of census coverage error, the Census Bureau should consider:

- **Use of several approaches before focusing on a specific model; besides logistic regression, alternatives should include use of other link functions, discriminant analysis, and various data mining approaches, such as classification trees, support vector machines, and neural nets.**
- **Thorough examination of the subset of predictors that is best suited to each individual statistical model; the predictors for these various statistical models need not be identical; however, there may be a benefit to constraining the (logistic regression) models of match rate and correct enumeration rate to have identical variables in the estimation of net coverage error, and research should be carried out to assess whether this benefit outweighs the benefit of selecting variables that are optimal for each of these two logistic regression models.**
- **To effectively blend information from auxiliary sources at various levels of geographic and demographic aggregation, random effects modeling and Bayes' methods also should be examined.**

Missing Data

In addition to developing a better understanding of the correlates of component census coverage error, a related issue that is worth mentioning is the development of a better understanding of the correlates of missing data—obviously related to the quality of the census—that may also be obtained through the use of statistical models. There are several different types of missing data in DSE. First, there are incomplete E-sample interviews, where the degree of "missingness" can range from characteristics information (e.g., missing demographic information) to situations in which the number of residents in a housing unit is unknown. Census interviews can also have missing information about whether a housing unit is occupied or whether it is even a housing unit. (There are also incomplete P-sample interviews, but given the more intensive field work, these are relatively infrequent.)

Second, missing characteristics information can affect whether E- and P-sample enumerations are able to be identified as matching or non-matching, whether a P-sample enumeration is properly designated as a resident of the housing unit or not, and whether an E-sample enumeration is correct or incorrect.

Third, missing characteristics information can affect either the use of poststrata or other means for smoothing or modeling estimates of the match rate or the correct enumeration rate.

Because missing data reduce the quality of census counts in various ways, it will be important for the Census Bureau to study which operational factors relate to its occurrence as a first step in trying to reduce its frequency in the future. Since missing data can be further categorized as occurring at the level of an entire household or an individual, useful dependent variables in such a study would be: (1) an indicator variable for missing household count, (2) various indicator variables either for specific missing household characteristics or for the degree of missing household characteristics, and (3) various indicator variables either for specific missing individual characteristics or for the degree of missing individual characteristics. Given these dependent variables, one could develop statistical models that would help identify which factors helped to discriminate between households and people with and without missing data. These models have the potential to make use of the entire census database, rather than just the data from the coverage measurement program.

In addition, given that proxy enumerations can be viewed as informed imputations, one might also develop statistical models that predict whether an enumeration is likely to be by proxy, again to assess which factors may be related and potentially causal. Since proxy enumerations are actual responses and not missing data, they can directly cause component coverage errors and therefore one could, as mentioned above, use such indicator variables as predictors in the study of causes of the rates of duplication, omission, erroneous enumeration, and enumeration in the wrong location.

Finally, PES data, through the matching of the P-sample and the E-sample, could be used to assess the frequency of failure of characteristic values to agree, such as race/ethnicity, age, or sex. Such errors can affect some count tabulations. Administrative records could also be used to assist in such studies. (Of course, some disagreements are due to imputations, but others are due to misresponse.)

CENSUS DATA PRODUCTS FOR PROCESS IMPROVEMENT

As the panel emphasizes throughout this report, the new objective for coverage measurement in 2010—to improve census processes through collecting and analyzing relevant information on census coverage error and its covariates—is challenging. There are many possible routes to enumerating a person in the census (or attempting to do so), and the propensity for coverage error is a complex function of both the census processes used to enumerate individuals and households and their characteristics. Even so, we are optimistic that much can be learned about which processes lead to greater frequencies of coverage error for individuals with certain characteristics and living situations. We believe that statistical analysis of data from the census coverage measurement program as designed for the 2010 census, possibly along with data from administrative records and the ACS, can provide considerable evidence about the causes of coverage error or at least suggest valuable hypotheses for further investigation.

How this information is summarized for data users, and how it is analyzed by people both inside and outside the Census Bureau has not yet been decided. This is an extremely important dimension of the coverage measurement program that needs to receive additional attention on the part of the Census Bureau.

As described above, the coverage measurement program in 2010 will provide information not only on net coverage error, but also on the four components of census coverage error.[2] With regard to net coverage error, somewhat analogous to 2000, the Census Bureau has proposed that it will

[2]There will be inconsistencies between the tabulations for net coverage error and for components of census coverage error that need to be communicated to census data users. In particular, while net census coverage error uses DSE to estimate those missed by both the census and the CCM, since those omissions have no information locating them in a housing unit, they are excluded from tabulations of components of coverage error and associated analyses.

produce estimates of the net coverage error nationally, for major demographic groups, and for states.

Although there are likely to be no poststrata in 2010 due to the use of logistic regression modeling, for ease of comparison the Census Bureau should consider releasing estimates of net coverage error for the 2010 census for comparable aggregates to support the comparison of net coverage error from 2000 to 2010.

For housing units, the Census Bureau will produce estimates of net coverage error nationally and by occupied status and other characteristics.

With regard to components of census coverage error, the Census Bureau proposes to produce the following research tables for persons: erroneous enumerations (by which is meant duplications and what is defined here as erroneous enumerations), and omissions nationally and by major demographic group, and erroneous enumerations by major census process. For housing units, the Census Bureau proposes to produce research tables on the rate of erroneous enumerations by major census process. In addition, the Census Bureau also proposes to produce research tables providing some assessment of the extent to which people are counted in the wrong geographic locations and the distances from the correct locations.

The panel commends the Census Bureau for its forward thinking about the role of coverage measurement in 2010 and the production of these various research tabulations. In particular, providing rates of erroneous enumerations by census process are likely to provide very useful information on the sources of census coverage errors. Also, given that measurements of the components of census coverage error have never been tabulated before, the public release of these tabulations for the 2010 census is especially to be commended, since the newness of these statistics argues for a somewhat conservative approach to their public release. Although census data users may not be asking about rates of components of census coverage error, the release of these tabulations will give data users a better sense of the totality of error in the census counts.

However, the panel believes that the Census Bureau needs to go further than the release of these tabulations, at least internally, in order to support the ultimate goal of a feedback loop that uses information from coverage measurement for identification of deficient census processes (and possibly even hints at preferred alternative processes). We believe that the major benefit from the current coverage measurement program could very well be the production of a database that provides *linked* information for purposes of analysis on: (1) persons and their households and housing units, (2) the census processes that were used to (try to) enumerate each individual or housing unit, (3) if an enumeration error was made, what type of error was made, and (4) possibly, contextual information from the relevant geographic area. Such a database could then support the use of regression-type models that would identify the covariates that were (and were not) predictive for rates of the four components of census coverage error.

To be more specific, an analytic database, based on 2010 CCM data plus auxiliary data, would contain important information. First, it would have descriptive information about each individual from either the P-sample interview or the E-sample interview, along with information on the housing unit (e.g., type of dwelling unit and owner/renter). Second, it would have the results of the census coverage measurement program assessment of whether the individual was properly counted, omitted, erroneously included, duplicated, or included at the wrong location. Also included would be the coverage status of other members of the housing unit (e.g., to establish when a person was an omission in an otherwise enumerated household and when the entire housing unit was missed).

Third, to the extent possible, this database would include the history of census processes that were used to attempt to enumerate each individual and household. This would include how the address was added to the MAF, whether the questionnaire was returned, whether the enumeration was obtained in nonresponse follow-up, etc. The panel has been told that for some processes this history is not retained. For instance, if someone was a duplicate and removed from the census during the follow-up process, the fact that a duplicate enumeration existed for a short period of time will likely not be retained. Especially given the critical importance of coverage follow-up and the need to evaluate its performance, the Census Bureau should consider retaining this history in a more complete form for just the CCM block clusters, i.e., essentially a master trace sample (for details, see National Research Council, 2004a).

StARS might also provide very helpful information on verifying whether a census coverage error was made, and so efforts should be made to fold in the StARS database. And the American Community Survey might provide contextual information at the small-area level, such as the poverty or unemployment rate. In addition, the quality of the enumerator work might be weakly assessed through information on enumerator turnover rates and other measures of the quality of the census field staff, based on information from the various management information systems for census enumeration districts. One might even argue for linking in the results of the quality assurance findings on the initial workload for each enumerator (which could be used to see whether the current decision rule of which enumerators to retain and which to let go is set at the right level).

The degree to which this database can and will be used to address important questions about census design is directly a function of the extent to which the needed analyses are permitted (even better, facilitated) by the structure of the database. This is a question of database management, which this panel did not address. However, it is clear that the development of such a database will not be easy, and it is therefore important that it be done by those with considerable expertise. The panel believes it is very likely that the Census Bureau would benefit from external assistance in developing such a database if it decides to pursue its development.

To indicate just one type of complication in creating such a database, consider the different geographies that might be encountered. There is the geography of the contextual information, which is likely to be census tracts. Then there is the geography of the CCM survey, which is the P-sample block clusters (though estimates will be available for any domain). There is also the geography of census enumeration districts. This variety of "geographies" complicates linkages of data.

The specification of the structure of such a database should start with the development of a list of the primary types of questions that the database will likely be asked to address. To this end, the coverage measurement staff should formulate various hypotheses that it would like a database to help resolve. Examples include: How much duplication is the result of children living in joint custody arrangements? How many erroneous enumerations are caused by the misinterpretation of residence rules? Could StARS be used to reduce the number of MAF omissions? The development of such a list will greatly assist in determining the structure of the database, including what information should be linked in, and how it should be linked.

In addition to using this database to confirm (or reject) a set of primary hypotheses of interest, such a database could also be used in a more exploratory manner to search for interesting and unanticipated patterns that lead to census coverage error. Every census seems to have surprises involving the types of people or households that experience novel causes of coverage error. A caution, though, is that it is the nature of such exploratory analyses to "discover" situations that are illusory or patterns that are idiosyncratic aspects of the 2010 census.

Recommendation 11: The primary output of the Census Bureau's coverage measurement program in 2010 should be an analytic database that is used to support the development of statistical models to inform census process improvement. The production of summary tabulations should be of lesser priority.

Having argued for the development and use of this database, it is the panel's responsibility to suggest how the analyses generated by this database could be used to support a feedback loop for census improvement. Our view is that for each hypothesis of interest, one could develop a statistical model to examine its validity. The development of such a model would likely be nontrivial, potentially involving conditioning on variables (such as sex of the individual, or urban/rural), and requiring diagnostic tools for identifying which predictors are useful and which are not, when transformations are and are not useful, etc. As mentioned above, many of these hypotheses will be modeled as forms of discriminant analyses, i.e., methods for using the predictors to separate the cases that were subject to a type of coverage error from those that were not.

When successful, such an analysis would suggest that certain types of housing units, or households, or individuals, were more often missed (or duplicated or otherwise incorrectly "counted") when associated with a particular census enumeration process. Although this information would be very useful, it would be incomplete, not providing the Census Bureau with what it needs most: an alternative process that would remedy the situation. For example, suppose an analysis demonstrated that many "Be Counted" enumerations were duplicates, by noticing that a model of the frequency of census duplication had a significant coefficient for the predictor that was an indicator variable for enumeration using the "Be Counted" program. (Such a finding would hardly be a surprise given the nature of the program.) The difficulty is that it might not be clear what could be done to the "Be Counted" program that would reduce the frequency of duplication. Is it the questionnaire itself that is being misunderstood? Is the problem where the "Be Counted" forms are distributed? Or is the problem how the forms are validated once they are collected? So it is important to realize that the full feedback loop will still require insight and interaction with field staff to use the information provided to identify what specifically should be modified to provide needed improvement in subsequent censuses.

Finally, we caution that such analyses cannot be used to tease out extremely detailed aspects of census taking. The size of the CCM, and the general infrequency of the four components of census coverage error (in 2000, there were 37 million estimated omissions and overcounts of 283 million enumerations and as discussed often in this report, that is an overestimate of the gross error rate; see National Research Council, 2004b) crossed by census processes used, among other variables, will necessarily limit the value of the proposed database to address very detailed questions. Consequently, the panel cautions that Herculean efforts to attempt to perfectly represent the census processes used through complicated indicator variables is probably not going to be profitable. Rather, general patterns of the census processes that proved to be less than fully effective in enumerating relatively substantial subpopulations is what one hopes to be able to identify. Such analytic findings may also be strengthened if they are supported by anecdotal reports of field staff that suggest causal explanations for the relationships discovered.

A remaining question is what should the Census Bureau communicate to census data users about the analyses that are produced by the proposed database? First, it is our hope that some extracts of this database could be made available, possibly through the Census Bureau data centers (taking care with respect to avoiding disclosure), to experts in various areas of statistical methodology to see if different patterns of interest can be identified. Second, while most census data users will be uninterested in either developing their own models or even the models that the Census Bureau develops, there is going to be interest in why the Census Bureau is making various changes in the census coverage measurement program: therefore, the Census Bureau may find it useful to issue a research report series on major findings from the analyses (both internally and externally) of a CCM database. These reports can describe the statistical models developed and report on improvements to the census design or census processes that might be indicated.

Recommendation 12: The Census Bureau should develop regression models that elucidate the various types of census coverage error, using specified dependent and predictor variables. To the extent that the database supporting these models can be made available to external researchers, it is extremely important that the Census Bureau pursue all viable avenues to involve outside researchers in the development of such models.

Given that the goal of the coverage measurement program in 2010 is census improvement, complete documentation of the functioning of the various census processes in 2010, at least for a sample of households, is crucially important to provide a baseline for comparison of potential alternative processes. Such a sample of households is commonly called a master trace sample. The value of such a master trace sample would be significantly increased if it substantially overlaps with the CCM sample, in order to be able to link indications of components of census coverage error with the census processes involved.

Recommendation 13: For a sample of households, the Census Bureau should retain data that provide a comprehensive picture of the census processes used to enumerate it, and the individuals residing in it, to facilitate subsequent evaluation. To allow linking assessment of census coverage error with a history of the census processes, this sample should substantially overlap with the CCM sample.

References and Bibliography

Adams, T., and Xijan, L. (2001). *Evaluation of Lack of Balance and Geographic Errors Affecting Person Estimates.* Report 2. Executive Steering Committee for A.C.E. Policy II (October 11), Decennial Statistical Studies Division. Washington, DC: U.S. Census Bureau.

Alho, J.M. (1990). Logistic regression in capture-recapture models. *Biometrics* 46:623-635.

Alho, J.M. (1994). Analysis of sample based capture-recapture experiments. *Journal of Official Statistics* 10(3):245-256.

Alho, J.M., and Spencer, B.D. (1997). The practical specification of the expected error of population forecasts. *Journal of Official Statistics* 13(3):203-225.

Alho, J.M., and Spencer, B.D. (2005). Statistical demography and forecasting. New York: Springer.

Alho, J.M., Mulry, M.H., Wurdeman, K., and Kim, J. (1993). Estimating heterogeneity in the probabilities of enumeration for dual-systems estimation. *Journal of the American Statistical Association* 88(423):1,130-1,136.

Anderson, M.J. (2000). *Encyclopedia of the U.S. Census.* M.J. Anderson, Editor-in-chief. Washington, DC: CQ Press.

Auer, P.L. (2004). *Study for Cases with Insufficient Information for Matching and Follow-up.* Unpublished paper (December 15). Washington, DC: U.S. Census Bureau.

Bates, N.A., and Gerber, E. (1998). Temporary mobility and reporting of usual residence. *Survey Methodology* 24(1):89-98.

Bell, W.R. (1993). Using information from demographic analysis in post-enumeration survey estimation. *Journal of the American Statistical Association* 88(423):1,106-1,118.

Bell, W.R. (2001). *Estimation of Correlation Bias in 2000 A.C.E. Using Revised Demographic Analysis Results.* Report 10. Executive Steering Committee for A.C.E. Policy II (October 16), Methodology and Standards Directorate. Washington, DC: U.S. Census Bureau.

Bell, W.R. (2005). *Contamination Between Census Coverage Follow-Up and Coverage Measurement—Considerations for Implementing Dual-Systems Estimation.* Unpublished paper (July 14). Washington, DC: U.S. Census Bureau.

Bell, W.R., Gibson, C.J., Das Gupta, P., Spencer, G.K., Robinson, J.G., Mulry, M.M., Vacca, E.A., Fay, R.E., and Leggieri, C.A. (1996). Report of the Working Group on the Use of Demographic Analysis in Census 2000. Washington, DC: Bureau of the Census, U.S. Department of Commerce.

Bench, K. (2002). *A Contamination of Census 2000 Data Collected in Accuracy and Coverage Evaluation Block Clusters.* Census 2000 Evaluation N.1 (August 22). Washington, DC: U.S. Census Bureau.

Brownrigg, L.A., and de la Puente, M. (1993). *Alternative Enumeration Methods and Results: Resolution and Resolved Population by Site.* 1990 Decennial Census Preliminary Research and Evaluation Memorandum No. 219. Washington, DC: U.S. Census Bureau. Available: http://www.census.gov/srd/papers/pdf/lab92-01.pdf [accessed 10/23/08].

Childers, D.R. (1992). The 1990 housing unit coverage study. *Proceedings of the Section on Survey Research Methodology of the American Statistical Association.* Available: www.amstat.org/sections/srms/Proceedings/y1992f.html [accessed 10/23/08].

Coale, A.J. (1955). The population of the United States in 1950 classified by age, sex and color—A revision of census figures. *Journal of the American Statistical Association* 50(269):16-54.

Coale, A.J., and Rives, N.W., Jr., (1972). *A Statistical Reconstruction of the Black Population of the United States, 1880-1970: Estimates of True Numbers by Age and Sex, Birth Rates, and Total Fertility.* Unpublished paper. Washington, DC : U.S. Census Bureau.

Coale, A.J., and Zelnik, M. (1963). *New Estimates of Fertility and Population in the United States.* Princeton, NJ: Princeton University Press.

Cowan, C.D., and Malec, D. (1986). Capture–recapture models when both sources have clustered observations. *Journal of the American Statistical Association,* (81):347–353.

de la Puenta, M. (2004). *Census 2000 Ethnographic Studies.* Census 2000 Topic Report No. 15 (March). Census Testing, Experimentation, and Evaluation Program. Washington, DC: U.S. Census Bureau.

Dillman, D.A., Treat, J.B., and Clark, J.R. (1994). Influence of 13 design factors on completion rates to decennial census questionnaires. 1994. *Annual Research Conference and CASIC Technologies Interchange Proceedings,* Rosslyn, Virginia, March.

DuMouchel, W.H., and Duncan, G.J. (1983). Using sample survey weights in multiple regression analyses of stratified samples. *Journal of the American Statistical Association* 78(383):535-543.

Ellis, Y. (1995). Examination of census omission and erroneous enumeration. Based on 1990 ethnographic studies of census coverage. *Proceedings of the Section on Survey Research Methods of the American Statistical Association.* Available: http://www.amstat.org/Sections/Srms/Proceedings/papers/1995_088.pdf [accessed 10/23/08].

Ericksen, E., Estrada, L.F., Tukey, J.W., and Wolter, K.M. (1991). *Report on the 1990 Decennial Census and the Post-Enumeration Survey.* Unpublished paper prepared for the Secretary. Washington, DC: U.S. Department of Commerce.

Fay, R.E. (2002). Probabilistic models for detecting census person duplication. *Proceedings of the Section on Survey Research Methods of the American Statistical Association.* Available: https://www.amstat.org/sections/SRMS/Proceedings/y2002/Files/JSM2002-000746.pdf [accessed 10/23/08].

Fay, R.E. (2003). Probabilistic models for detecting census duplication at the person and household levels. *Proceedings of the Section on Government Statistics of the American Statistical Association.* Available: https://www.amstat.org/sections/SRMS/Proceedings/y2002/Files/JSM2002-000746.pdf [accessed 11/20/08].

Fay, R.E. (2004). An analysis of person duplication in census 2000. *Proceedings of the Section on Survey Research Methods of the American Statistical Association.* Available: http://www.amstat.org/sections/srms/proceedings/y2004f.html [accessed 10/23/08].

Fein, D.J. (1990). Racial and ethnic differences in U.S. census omission rates. *Demography* 27(2):285-302.

Feldpausch, R. (2001). *Census Person Duplication and the Corresponding A.C.E. Enumeration Status. Report 6.* Executive Steering Committee for A.C.E. Policy II (October 31), Decennial Statistical Studies Division. Washington, DC: U.S. Census Bureau.

Fenstermaker, D.A. (2005). *2010 Census Coverage Measurement: Provisional Sample Allocation.* DSSD 2010 Census Coverage Measurement Memorandum Series #2010-C-01 (October 26). (Prepared by Peter P. Davis.) Washington, DC: U.S. Census Bureau.

Fenstermaker, D.A. (2006). *2010 Census Coverage Measurement: Methodology and Results for the Provisional Sample Allocation.* DSSD 2010 Census Coverage Measurement Memorandum Series #2010-C-xx (April 2006–Draft). (Prepared by Peter P. Davis and Tom Mule.) Washington, DC: U.S. Census Bureau.

Fenstermaker, D.A., and Mule, T. (2002). *A.C.E. Revision II Results: Further Study of Person Duplication.* DSSD A.C.E. Revision II Memorandum Series #PP-51 (December 31). Washington, DC: U.S. Census Bureau. Available: http://www.census.gov/dmd/www/pdf/pp-30r-5.pdf [accessed 10/23/08].

Fowlkes, E.B. (1987). Some diagnostics for binary logistic regression via smoothing, *Biometrika* 74(3):503-515.

Garcia-Parra, C. (1987). *Causes for Undercoverage in East Los Angeles from an Anthropological Perspective.* Report prepared under contract, April 1987, Applied Anthropology Documentation Project (unknown binding).

Griffin, R.A. (2000). Post-stratification classification affected by household coverage. *Proceedings of the Survey Research Methods Section of the American Statistical Association.* Available: http://www.amstat.org/sections/SRMS/proceedings/papers/2000_081.pdf [accessed 10/23/08].

Griffin, R.A. (2005). *Preliminary Research on Separate Estimation of the Population in Enumerated and Non-Enumerated Housing Units.* DSSD 2010 Census Coverage Measurement Memorandum Series # 2010-E06 (November 29). Washington, DC: U.S. Census Bureau. (prepared by Thomas Mule).

Griffin, R.A, Mule, T., and Olson, D. (2005). *Initial Results of Preliminary Net Error Empirical Research Using Logistic Regression.* DSSD 2010 Census Coverage Measurement Memorandum Series # 2010-E03 (April 18). Washington, DC: U.S. Census Bureau. Available: http://www.amstat.org/sections/srms/Proceedings/y2006f.html [accessed 10/23/08].

Habermann, S.J., Jiang, W., and Spencer, B.D. (1998). *Activity 7: Develop Methodology for Evaluating Model-Based Estimates of the Population Size for States.* Unpublished paper. Washington, DC: U.S. Census Bureau.

Hawley, W. (2004). *MAF/TIGER Accuracy Improvement Project—An Update.* PowerPoint Presentation to WaTrans Partner's Meeting (March 2). Available: http://www.wsdot.wa.gov/mapsdata/transframework/project_presentations/wa-trans_mtaip_update_03-02-04.pdf [accessed 10/24/08].

Hill, J.M., and Machowski, J.D. (2003). *Census 2000 Evaluation B.6., Master Trace Sample.* Unpublished paper (September 29), Planning, Research, and Evaluation Division. Washington, DC: U.S. Census Bureau.

Hogan, H. (2003). The Accuracy and Coverage Evaluation: Theory and Design. *Survey Methodology,* 29(2):129-138.

Huggins, R.M. (1989). On the statistical analysis of capture-recapture experiments. *Biometrika* 76:133-140.

Jaros, Stephanie. (2007). *Master Bibliography on Ethnography and the Census.* Unpublished paper. Committee on National Statistics.

Kadane, J.B., Meyer, M.M., and Tukey, J.W. (1999). Yule's association paradox and ignored stratum heterogeneity in capture-recapture studies. *Journal of the American Statistical Association* 94:855-859.

Kearney, A.T. (2002). *A.C.E. Revision II Missing Data Evaluation Final Report.* DSSD A.C.E. Revision II Memorandum Series #PP-48. Washington, DC: U.S. Census Bureau. Available: http://www.census.gov/dmd/www/pdf/pp-48r.pdf [accessed 10/23/08].

Keller, A. (2006). Inconsistency between Accuracy and Coverage Evaluation Revision II and demographic analysis estimates for children 0-9 years of age. *Proceedings of the Survey Research Methods Section of the American Statistical Association.* Available: http://www.amstat.org/sections/srms/Proceedings/y2006f.html [accessed 11/20/08].

Kostanich, D., and Whitford, D. (2005). *Discussion of the Proposal to Reduce Respondent Burden for Households in Both Census Coverage Follow-up and Census Coverage Measurement.* DSSD 2010 Census Coverage Measurement Memorandum Series #A-07 (November 29). Washington, DC: U.S. Census Bureau.

Lee, R.D., and Tuljapurkar, S.D. (1994). Stochastic population forecasts for the United States: Beyond high, medium, and low. *Journal of the American Statistical Association* 89:1,175-1,189.

Little, R.J.A. (2003). *To Model or Not to Model? Competing Modes of Inference for Finite Population Sampling.* Department of Biostatistics Working Paper Series, Working Paper 4, University of Michigan.

Little, R.J.A., and Rubin, D.B. (2002). *Statistical Analysis with Missing Data.* 2nd Edition. New York: John Wiley.

Malec, D., and Maples, J. (2005). An evaluation of synthetic, small-area census coverage error using a random effects model. *Proceedings of the Section on Survey Research Methods of the American Statistical Association.* Available: http://www.amstat.org/Sections/Srms/Proceedings/y2005/Files/JSM2005-000180.pdf [accessed 11/20/08].

Mallows, C.L. (1973). Some comments on Cp. *Technometrics* 15:661-675.

Marks, E.S., Seltzer, W., and Krotki, K.J. (1974). *Population Growth Estimation.* New York: The Population Council.

Martin, E. (1999). Who knows who lives here? Within-household disagreements as a source of survey coverage error. *The Public Opinion Quarterly* 63(2):220-237.

Martin, E.A., Brownrigg, L.A., and Fay, R.E. (1990). Results of the 1988 ethnographic studies of census coverage and plans for 1990. *Proceedings of the Section on Survey Research Methods of the American Statistical Association.* Available: https://www.amstat.org/sections/srms/Proceedings/papers/1990_089.pdf [accessed 10/23/08].

Moriarity, C., and Childers, D. (1993). Analysis of census omissions: Preliminary results. *Proceedings of the Section on Survey Research Methods of the American Statistical Association.* Available: http://www.amstat.org/sections/srms/Proceedings/papers/1993_104.pdf [accessed 10/23/08].

Mule, T. (2001). *ESCAP II Person Duplication in Census 2000.* Executive Steering Committee for A.C.E. Policy II, Report 20 (October 11). Decennial Statistical Studies Division. Washington, DC: U.S. Census Bureau.

Mule, T. (2003). *A.C.E. Revision II Results: Change in Estimated Net Undercount.* DSSD A.C.E. Revision II Memorandum Series PP-58. Washington, DC: U.S. Census Bureau.

Mule, T. (2007). Coverage measurement for the 2010 census. Presented at the Washington Statistical Society Meeting, Washington, DC, January 23, 2008. Available: http://scs.gmu.edu/~wss/wss080123slides.pdf [accessed 11/25/08].

Mule, T., Schellhammer, T., Malec, D., and Maples, J. (2007). *Using Continuous Variables as Modeling Covariates for Net Coverage Estimation.* DSSD 2010 Census Coverage Measurement Memorandum Series #2010-E-09-RI (March 6). Washington, DC: U.S. Census Bureau.

Mulry, M.H., and Kostanich, D.K. (2006). Framework for census coverage error components. *Proceedings of the Section on Survey Research Methods of the American Statistical Association.* Available: http://www.amstat.org/sections/srms/Proceedings/y2006f.html [accessed 10/23/08].

Mulry, M.H., and Spencer, B.D. (1991) Total error is PES estimates of population: The dress rehearsal census of 1998. *Journal of the American Statistical Association*, 86:839-854.

Mulry, M.H., Schindler, E., Mule, T., Nguyen, N., Spencer, B.D. (2005). Investigation of extreme estimates of census coverage error for small areas. *Proceedings of the Section on Survey Research Methods of the American Statistical Association*. Available: http://www.amstat.org/Sections/Srms/Proceedings/y2005/Files/JSM2005-000551.pdf [accessed 10/23/08].

National Research Council. (1985). *Bicentennial Census: New Directions for Methodology in 1990*. Panel on Decennial Census Methodology, Constance F. Cirtro and Michael L. Cohen (Eds.). Committee on National Statistics, Commission of Behavioral and Social Sciences. Washington, DC: National Academy Press.

National Research Council. (2000). *Redesigning the 2010 Census: First Interim Report*. Panel on Research on Future Census Methods, Committee on National Statistics. Washington, DC: National Academy Press.

National Research Council. (2004a). *Reengineering the 2010 Census: Risks and Challenges*. Panel on Research on Future Census Methods. Daniel L. Cork, Michael L. Cohen, and Benjamin F. King (Eds.). Committee on National Statistics, Division of Behavioral and Social Sciences and Education. Washington, DC: The National Academies Press.

National Research Council. (2004b). *The 2000 Census: Counting Under Adversity*. Panel to Review the 2000 Census, Constance F. Citro, Daniel L. Cork, and Janet L. Norwood (Eds.). Committee on National Statistics, Division of Behavioral and Social Sciences and Education. Washington, DC: The National Academies Press.

National Research Council. (2006). *Once, Only Once, and in the Right Place: Residence Rules in the Decennial Census*. Panel on Residence Rules in the Decennial Census, Daniel L. Cork and Paul R. Voss, Eds. Committee on National Statistics, Division of Behavioral and Social Sciences and Education. Washington, DC: The National Academies Press.

National Research Council. (2007). *Research and Plans for Coverage Measurement in the 2010 Census: Interim Assessment*. Panel on Correlation Bias and Coverage Measurement in the 2010 Decennial Census, Robert Bell and Michael L. Cohen (Eds.). Committee on National Statistics. Division of Behavioral and Social Sciences and Education. Washington, DC: The National Academies Press.

Obenski, S., Farber J., and Chappell, G. (2005). Research to Improve Census Imputation Methods: Item Results and Conclusions. *Proceedings of the 2005 Joint Statistical Meetings*, Survey Research Methods Section, Alexandria, VA: American Statistical Association.

Passel, J.S. (2005). *Estimates of the Size and Characteristics of the Undocumented Population*. Pew Hispanic Center: Washington, DC. March. Available: http://pewhispanic.org/reports/report.php?ReportID=44 [accessed 11/2/08].

Passel, J.S., Word, D.L., McKenney, N.D. and Kim, Y. (1982). *Postcensal estimates of the Asian population in the United States: description of methods using surname and administrative records*. Paper presented at the annual meeting of the Population Association of America, San Diego, California, April 1982.

Poole, D., and Raftery, A.E. (2000). Inference for deterministic simulation models: The Bayesian melding approach. *Journal of the American Statistical Association* 95(452):1,244-1,255.

Price, D.O. (1947). A check on underenumeration in the 1940 census. *American Sociological Review* 12 (1):44-49.

Raghunathan, T.E., P.W. Solenberger, and J. Van Hoewyk. (2006). IVEware: Imputation and Variance Estimation Software. Survey Methodology Program, Survey Research Center, Institute for Social Research, University of Michigan. Available: http://www.isr.umich.edu/src/smp/ive/ [accessed 21/12/06].

Resnick, D., Love, S., Taeuber, C., and Staveley, J.M. (2004). *Analysis of ACS Food Stamp Program participation underestimate*. Paper presented at the 2004 Joint Statistical Meeting, Toronto.

Robinson, J.G. (2001). *Demographic Analysis Results.* Executive Steering Committee for A.C.E. Policy II, Report No. 1, October 13, 2001. Washington, DC: U.S. Census Bureau.

Robinson, J.G., and West, K.K. (2000). *Demographic Analysis,* pp. 165-168. Encyclopedia of the U.S. Census. M.J. Anderson, Editor-in-chief. Washington, DC: CQ Press.

Robinson, J.G., Ahmed, B., Das Gupta, P., and Woodrow, K.A. (1993). Estimation of population coverage in the 1990 United States census based on demographic analysis. *Journal of the American Statistical Association* 88(423):1,061-1,071.

Rubin, D.C. (1988). Using the SIR algorithm to simulate posterior distributions. Pp. 395-402 in *Bayesian Statistics 3,* J.M. Bernardo, M.H. DeGroot, D.V. Lindley, and A.F. Smith (Eds.). Oxford, UK: Clarendon Press.

Ruhnke, M.C. (2003). *An Assessment of Addresses on the Master Address File Missing in the Census or Geocoded to the Wrong Collection Block.* Census 2000 Evaluation F.15. Washington, DC: U.S. Census Bureau.

Schindler, E. (2006). 2010 census coverage measurement: The hunt for the magic variables. *Proceedings of the Section on Survey Research Methods of the American Statistical Association.* Available: http://www.amstat.org/sections/srms/Proceedings/y2006f.html [accessed 10/23/08].

Sekar, C.C., and Deming, W.E. (1949). On a method of estimating birth and death rates and the extent of registration. *Journal of the American Statistical Association* 44: 101-115.

Shoemaker, H.H. (2005). *Results of a Feasibility Study to Match Census Enumerations Coded in A.C.E. as Insufficient for Matching and Follow-up.* DSSD 2010 Census Coverage Measurement Memorandum Series #2010-B01 (April 18). Washington, DC: U.S. Census Bureau.

Shores, R. (2002). *A.C.E. Revision II Adjustment for Correlation Bias.* DSSD A.C.E. Revision II Memorandum Series PP-53, (December 31). Washington, DC: U.S. Census Bureau. Available: http://www.census.gov/dmd/www/pdf/pp-53r.pdf [accessed 10/23/08].

Shryock, H. and Siegel, J.S. (1973). The methods and materials of demography, Part I. Washington, DC: U.S. Department of Commerce.

Siegel, J.S. (1974). Estimates of coverage of the population by sex, race, and age in the 1970 census. *Demography* 11(1):1-23.

Siegel, J.S., and Zelnik, M. (1966). An evaluation of coverage in the 1960 census of population by techniques of demographic analysis and by composite methods. *Proceedings of the Social Statistics Section of the American Statistical Association,* pp. 71-85. Washington, DC: American Statistical Association.

Spencer, B.D. (2005). *Statistical Analysis Research on Small Area Estimation of Census Coverage.* Unpublished paper prepared by Abt Associates Inc., and Spencer Statistics, Inc. (March 22). Washington, DC: U.S. Census Bureau.

Stuart, E.A., and Zaslavsky, A.M. (2002). Using administration records to predict census day residency. Pp. 335-349 in *Case Studies in Bayesian Statistics, Volume VI,* C. Gatsonis, R.E. Kass, A. Carriquiry, A. Gelman, D. Higdon, D.K. Pauler, and I. Verdinelli (Eds.). New York: Springer.

Trussell, J. (1981). Should state and local area census counts be adjusted? *Population Index* 47(1):4-12.

Tukey, J.W. (1983). Affidavit presented to District Court, Southern District of New York, Mario Cuomo, et al., Plaintiff(s), Malcom Baldridge, et al., Defendants, 80 Civ. 4550 (JES).

U.S. Census Bureau. (2003). *Technical Assessment of A.C.E. Revision II.* Unpublished paper (March 12). Washington, DC: Author.

U.S. Census Bureau. (2004a). *Accuracy and Coverage Evaluation of Census 2000: Design and Methodology.* Unpublished paper. Washington, DC: Author.

U.S. Census Bureau. (2004b). *The 2006 Census Coverage Measurement Test.* Unpublished draft paper (April 7). Washington, DC: Author.

Valentine, C.B., and Valentine, B.L. (1971). *Missing Men: A Comparative Methodological Study of Underenumeration and Related Problems*. Report prepared under Joint Statistical Agreement with the Brookdale Hospital Center. Washington, DC: U.S. Census Bureau.

Wang, X., Lim, J., and Stokes, S.L. (2006). Forming post-strata via Bayesian treed capture-recapture models. *Biometrika* 93:861-876.

Zaslavsky, A.M., and Wolfgang, G.S. (1990). Triple system modeling of Census, Post-Enumeration Survey, and Administrative List Data. *American Statistical Association Proceedings of the Section on Survey Research Methods* 668-673.

Appendix A

A Framework for Components of Census Coverage Error

This appendix summarizes Mulry and Kostanich (2006). They begin by hypothesizing a P-census, which is the P-sample if the entire United States were included in a postenumeration survey (PES). The P-census is also idealized in that no errors are assumed to be made in its data collection or matching, though the P-census can miss, at random, some correct enumerations in the census.

The authors then categorize people on the basis of the quality of their data, that is, whether their census questionnaire has errors or nonresponse, as follows:

1. those correctly enumerated in the census, *CE*,
2. those enumerated in the census but in the wrong location, *WL*,
3. those erroneously enumerated in the census, *EE*,
4. those with insufficient information for matching to the P-census, *II*,
5. those that are not data defined in the census, *NDD*, and
6. those omitted in the census, *OM*.

The authors also divide the population into four subsets by crossing the following two dichotomies: whether or not a census enumeration has sufficient information for matching and whether or not a census enumeration is in the P-census. The subscript *ij* indicates subset membership: the first index is equal to 1 for those with sufficient information for matching and 0 otherwise; the second index is equal to 1 with inclusion in the

<table>
<tr><td rowspan="2">Census</td><td rowspan="2">Eligible for E-sample</td><td rowspan="2">Matching Universe</td><td colspan="3">P-Census</td></tr>
<tr><td>In</td><td colspan="2">Not In</td></tr>
<tr><td rowspan="4">In</td><td rowspan="3">In</td><td rowspan="2">In</td><td>CE_{11}</td><td>CE_{10}</td><td rowspan="2">EE_{10}</td></tr>
<tr><td>WL_{11}</td><td>WL_{10}</td></tr>
<tr><td rowspan="3">Not In</td><td>II_{01}</td><td>II_{00}</td><td>$EEII_{00}$</td></tr>
<tr><td rowspan="2">Not In</td><td>NDD_{01}</td><td>NDD_{00}</td><td>$EENDD_{00}$</td></tr>
<tr><td>Not In</td><td>OM_{01}</td><td colspan="2">OM_{00}</td></tr>
</table>

FIGURE A-1 Elements of dual-systems estimation.
SOURCE: Adapted from Mulry and Kostanich (2006).

P-census and 0 otherwise. See Figure A-1 for a depiction of the various subsets of the total population using this taxonomy.

The result is 13 separate cells, defined as follows:

CE_{11}: correct enumeration in the census and in the P-census
CE_{10}: correct enumeration in the census and missed in the P-census
EE_{10}: erroneous enumeration in the census and missed in the P-census (which would include both erroneous enumerations as defined in this report and duplicate enumerations in the census
$EEII_{00}$: erroneous enumeration in the census with insufficient information for matching and missed in the P-census
$EENDD_{00}$: erroneous enumeration in the census and not data-defined and missed in the P-census
WL_{11}: enumerated in the wrong location in the census and in the P-census
WL_{10}: enumerated in the wrong location in the census and missed in the P-census
II_{01}: insufficient information for matching in the census and counted in the P-census
II_{00}: insufficient information for matching in the census and missed in the P-census
NDD_{01}: not data defined in the census and in the P-census
NDD_{00}: nor data defined in the census and missed in the P-census
OM_{01}: missed in the census and in the P-census
OM_{00}: missed in the census and missed in the P-census

The following additional relationships are used below:

$$CE = CE_{11} + CE_{10}$$
$$WL = WL_{11} + WL_{10}$$

$$II = II_{01} + II_{00} + EEII_{00}$$
$$NDD = NDD_{01} + NDD_{00} + EENDD_{00}$$
$$OM = OM_{01} + OM_{00}$$

Thus:

$$\begin{aligned}
&Census = \\
&\quad CE_{11} + CE_{10} + WL_{11} + WL_{10} + II_{01} + II_{00} + NDD_{01} + NDD_{00} + \\
&\quad EE_{10} + EEII_{00} + EENDD_{00}; \\
&True\ Population = \\
&\quad CE_{11} + CE_{10} + WL_{11} + WL_{10} + II_{01} + II_{00} + NDD_{01} + NDD_{00} + \\
&\quad OM_{01} + OM_{00}; \\
&Net\ Census\ Error = \\
&\quad True\ Population - Census = OM_{10} + OM_{00} - EE_{10} - \\
&\quad EEII_{00} - EENDD_{00}; \\
&P\text{-}Census = \\
&\quad CE_{11} + WL_{11} + II_{01} + NDD_{01} + OM_{01}
\end{aligned} \tag{1}$$

Given that the number of correct enumerations, CE, is equal to $CE_{11} + CE_{10}$; that the number of enumerations in the P-census, P, is equal to $CE_{11} + WL_{11} + II_{01} + NDD_{01} + OM_{01}$; and that the number of the P-census matches to correct census enumerations in the matching universe in the correct location, M, is equal to CE_{11}, one can re-express the dual-systems estimator,

$$DSE = CE\frac{P}{M},$$

in terms of the cell counts as

$$DSE = \left(CE_{11} + CE_{10}\right)\frac{\left(CE_{11} + WL_{11} + II_{01} + NDD_{01} + OM_{01}\right)}{CE_{11}}. \tag{2}$$

To justify this formula, the authors express three assumptions that are used in practical implementation of dual-systems estimation as a function of the entire set of 13 quantities:

Assumption 1: The basic assumption underlying dual-systems estimation is that the proportion of the true population correctly enumerated in the census equals the proportion of the P-census enumerated in the census.

This can be expressed as

$$\frac{CE + WL + II_{01} + II_{00} + NDD_{01} + NDD_{00}}{DSE} = \frac{CE_{11} + WL_{11} + II_{01} + NDD_{01}}{CE_{11} + WL_{11} + II_{01} + NDD_{01} + OM_{01}}.$$

Turning this around:

$$DSE = \left(CE + WL + II_{01} + II_{00} + NDD_{01} + NDD_{00}\right)\left(\frac{CE_{11} + WL_{11} + II_{01} + NDD_{01} + OM_{01}}{CE_{11} + WL_{11} + II_{01} + NDD_{01}}\right). \quad (3)$$

Assumption 2: It is assumed that correct enumerations in the matching universe are included in the P-census at the same rate as all correct enumerations.

That is, it is assumed that cases insufficient for matching can be treated as missing completely at random. This is expressible as

$$\frac{CE_{11} + WL_{11}}{CE + WL} = \frac{CE_{11} + WL_{11} + II_{01} + NDD_{01}}{CE + WL + II_{01} + II_{00} + NDD_{01} + NDD_{00}}. \quad (4)$$

Assumption 3: Given that the search for a match is geographically limited, it is assumed that the proportion of people that should be enumerated but are called erroneous because they are in the wrong location equals the proportion of matches that are not found because they are in the wrong location.

This assumption is the so-called balancing of erroneous enumerations and nonmatches and is equivalent to the statement that the proportion of correct enumerations found because they are in the correct location equals the percentage of matches found because they are in the correct location. This can be expressed as

$$\frac{CE_{11} + CE_{10}}{\left(CE_{11} + CE_{10} + WL_{11} + WL_{10}\right)} = \frac{CE_{11}}{CE_{11} + WL_{11}},$$

which can be re-expressed as

$$\frac{CE_{11} + CE_{10}}{CE_{11}} = \frac{\left(CE_{11} + CE_{10} + WL_{11} + WL_{10}\right)}{CE_{11} + WL_{11}} = \frac{CE + WL}{CE_{11} + WL_{11}}. \quad (5)$$

Substituting expressions (4) and (3) into (2), we have:

$$DSE = \left(CE_{11} + CE_{10}\right)\frac{\left(CE_{11} + WL_{11} + II_{01} + NDD_{01} + OM_{01}\right)}{CE_{11}}, \quad (6) = (2)$$

therefore justifying dual-systems estimation when the above three assumptions obtain.

The dual-systems estimation expression can be rewritten as

$$DSE = \left(CE_{11} + CE_{10} + WL_{11} + II_{01} + NDD_{01} + OM_{01}\right) + \frac{CE_{10}}{CE_{11}}\left(WL_{11} + II_{01} + NDD_{01} + OM_{01}\right),$$

which is equal to the true population if the last term is equal to the missing elements in expression (1): that is, if

$$\frac{CE_{10}}{CE_{11}}\left(WL_{11}+II_{01}+NDD_{01}+OM_{01}\right)=WL_{10}+II_{00}+NDD_{00}+OM_{00}. \quad (7)$$

The quantity on the right-hand side of (7) is referred to as the *fourth cell*—the people who are missed by both the census and by the P-census. If one assumes that the property of being correctly included in the census at the correct location is statistically independent of being in the P-census, then

$$\left[\frac{CE_{10}}{CE_{11}}\right]\left(CE_{11}+WL_{11}+II_{01}+NDD_{01}+OM_{01}\right)=CE_{10}+WL_{10}+II_{00}+NDD_{00}+OM_{00},$$

which is equivalent to (7).

Mulry and Kostanich also discuss what information is available from the field as to which of the sample of census enumerations, and which of the P-sample enumerations (many of which are the same individuals) fall into the various 13 types of enumerations listed above. Recall that the P-sample enumerations are only matched to matchable census enumerations in a search area. Also, for persons who have moved into the P-sample block clusters after census day, the P-sample is matched to their residence address on census day. Matches therefore provide an estimate of the number of correct enumerations in the correct location that were included in the P-sample. The P-sample is composed of matches and nonmatches: the matches, again ignoring sampling variation, are equal to CE_{11}, and the nonmatches are equal to $II_{01} + WL_{11} + NDD_{01} + OM_{01}$. These various types of nonmatches are not distinguishable without further data collection.

The number of census enumerations is the sum of the correct enumerations and erroneous enumerations (as defined by the Census Bureau), or $E = CE + EE$, where $CE = CE_{11} + CE_{10}$. In the expression $CE = CE_{11} + CE_{10}$, the components are distinguishable for nonmovers because in matching the P-sample to the E-sample, it is determined which census enumerations were included and which were missed in the P-sample. However, the two components of correct enumerations are not distinguishable for movers.

Mulry and Kostanich further address the measurement of components of census coverage error. If one wants to decompose the various summary estimates, more information would be needed than that used to support dual-systems estimation.

When the objective is the estimation of net coverage error, a very strict definition of correct enumeration is used, involving a small restricted search area within the relevant P-sample block cluster (and possibly a small area surrounding that area). But when the objective is to measure components of census coverage error, one can define a correct enumeration in a variety of ways to conform to a given tabulation of interest.

For instance, a correct enumeration can be in the correct county, state, or simply included correctly in the United States, the latter being the approach taken to simplify the argument given.

Mulry and Kostanich state their goal is partly to obtain estimates of the number of erroneous enumerations, $EE_{10} + EEII_{00} + EENDD_{00}$, and the number of census omissions, $OM_{01} + OM_{00}$. (In this report, the panel states there is also interest in estimating the number of enumerations in the wrong place and the number of duplicate enumerations.)

Unfortunately, because of enumerations in the wrong location and enumerations with either insufficient information for matching or not data defined, subtracting CE from the census count gives an inflated estimate of the number of erroneous enumerations, $EE_{10} + EEII_{00} + EENDD_{00}$. Specifically, $Census - CE_{11} - CE_{10} = WL_{11} + WL_{10} + II_{01} + II_{00} + NDD_{01} + NDD_{00} + EE_{10} + EEII_{00} + EENDD_{00}$, so $Census - CE$ is the sum of erroneous census enumerations (which includes duplicates) plus census enumerations in the wrong location plus correct census enumerations with insufficient information for matching.

For the same reason as for erroneous enumerations, subtracting the matching enumerations from the P-census does not provide an unbiased estimate of the number of omitted people in the census, $OM_{10} + OM_{00}$. In fact, $P - M = II_{01} + WL_{11} + NDD_{01} + OM_{01}$. To obtain an estimate of the number of omissions, note that $DSE - Census = NetCensusError = OM_{01} + OM_{00} - EENDD_{00} + EEII_{00} + EE_{10}$, and, therefore, $OM_{01} + OM_{00} = NetCensusError + EENDD_{00} + EEII_{00} + EE_{10}$. So, to estimate the number of omissions, one can take an estimate of the net census error and add to it the number of erroneous enumerations (including the number of duplicates).

The Census Bureau plans to use two definitions of a correct enumeration in 2010, one to provide a quality estimate of net census error, which among other things will help to estimate the number of omissions, and one to estimate the remaining components of coverage error.

To estimate the number of erroneous enumerations, the Census Bureau will need:

- to collect additional data to determine where enumerations should be included if the search area is not the correct location;
- to match the E-sample enumerations against the full set of census enumerations for duplicates, with field validation if necessary to establish proper census residence; and
- for enumerations in the E-sample but not in the matching universe, to strive to match to the P-sample (when possible) to identify those KEs (responses that are census data-defined but have insuf-

ficient information for matching as defined in 2000) that are correct enumerations.

This appendix omits the remaining details: Mulry and Kostanich discuss how one could separate out those enumerated in the wrong location from those that are erroneous, other complications raised by cases with insufficient information for matching, movers, and duplicates, and when to use imputation methods. Finally, the estimates of the components are generally represented as sample weighted averages, mainly of 0-1 indicator variables, but also of imputed probabilities.

Appendix B

Logistic Regression for Modeling Match and Correct Enumeration Rates

It is reasonable to suspect that match rates and correct enumeration rates, in addition to being a function of the variables used to define the accuracy and coverage evaluation (A.C.E.) poststrata in 2000, may also vary across the local census offices used to manage the workload in the census. The local office identifiers are on the A.C.E. research database, but they were not included in the six logistic regression models described above or the study by Schindler (2006).

Local census office indicator variables might be predictive of match and correct enumeration rates because factors that are particular to small areas could affect ease of enumeration. For example, local economic conditions and the expertise and capabilities of local census office administrators could vary. Because of the large number of local census offices (more than 500) and the limited amount of data for each, these effects are more naturally represented as random effects. By including these random effects in the logistic regression models, the Census Bureau could estimate the effects of individual offices on match and correct enumeration rates and obtain valid estimates of the contribution of variability across offices to uncertainty about coverage rates in each area.

Malec and Maples (2005) explored this approach by adding local area random effects into a synthetic estimation model and then measured the variance component of these random effects for local census offices. The ultimate objective of this approach is a small-area estimation methodology that would provide a compromise between synthetic estimation and a design-based estimator for each local office area.

Because of the complex design of A.C.E.'s postenumeration survey (weighted cases within samples of block clusters), many of the empirically correct enumeration rates and match rates used in Malec and Maple's model are more variable than the nominal sample sizes would indicate. To account for the extra variability, Malec and Maples (2005) used a pseudo-likelihood approach with effective sample sizes estimated by the bootstrap approach.

In this approach, both logistic regression models (for match rate and correct enumeration rate) have the following generic form:

$$\log\left(\frac{p_{i,k}}{1-p_{i,k}}\right) = \beta_i + \mu_k + \alpha_{i,k},$$

where β_i is the fixed effect for i^{th} poststratum membership, μ_k is a random effect for the kth local census office, and α_{ik} is model error. Furthermore,

$$\mu_k \sim N\left(\underline{0}, \Sigma\right) \text{ and } \alpha_{ik} \sim N\left(\underline{0}, \gamma^2_{ce(i)}\right),$$

where *ce(i)* is an index representing the collapsing of the poststrata into 11 or 8 cells, depending on whether the model is applied to the E-sample or the P-sample. Malec and Maples (2005) were able to estimate the large number of parameters in these models using Bayesian simulation.

This research suggests that inclusion of small-area effects could substantially improve coverage estimates. Several questions remain: how best to treat the complex sample design, how many random effects can be included and at what level of aggregation, the best way to estimate the model parameters, and how the model fit should be assessed. The panel is impressed with this high-caliber research that addresses an important issue in coverage modeling; further work in this area would be very valuable.

Mulry et al. (2005) examined the following anomalous results in A.C.E. More than 5 percent of incorporated places[1] in 2000 had an estimated net *overcount* of greater than 5 percent, and 0.5 percent had a net overcount of greater than 10 percent. This result runs counter to findings from the 1980 and 1990 coverage measurement programs of the potential net overcoverage due to true erroneous enumerations and duplications. In contrast with 2000, only 0.1 percent of places had an estimated net undercount of greater then 5 percent, and nationally, the degree of overcoverage and undercoverage were of essentially the same magnitude. There is a concern that the lack of balance of designated erroneous enumerations and designated omissions may be due to the use of proxy status and the type of census return as poststratification variables for the E-sample but not for P-sample computations.

[1]See http://www.census.gov/dmd/www/ACEREVII_PLACES.txt for a list.

To examine this further, Mulry et al. (2005) demonstrated that by using proxy status in the E-sample poststratification, there were 91 places with a net overcount of more than 10 percent: however, if it is assumed that there was no error for proxy enumerations, there were only 16 places with net overcounts of more than than 10 percent. Furthermore, if one assumes that there were no errors for proxy enumerations and no errors for late nonmail returns, there were only four places with a net overcount of more than 5 percent. Given this and given that 27 percent of proxy enumerations had insufficient information for matching and follow-up, it is clear that proxy enumerations could contribute to substantial balancing error. The Census Bureau concluded that proxy enumerations contributed to these anomalous findings, but that it was not the only cause.

Related research carried out by Spencer (2005) examined the quality of synthetic estimates for block clusters based on A.C.E. revision II estimates, either using 938 E-sample poststrata and 648 P-sample poststrata or using the same 648 poststrata for the E- and P-samples. His findings, in which the standard of comparison was either (a) the direct dual-systems estimate or (b) the census count plus people found in the P-sample who were omitted in the census for each block cluster, suggested that coarser but consistent poststrata may have provided more accurate estimates of net coverage error than finer poststratifications based on different E- and P-sample stratifications. However, for large blocks with proxy rates greater than 10 percent, the finer and inconsistent poststrata performed better.

The specific model form for logistic regression is

$$\log\left(\frac{p}{(1-p)}\right) = \boldsymbol{X}\beta.$$

As described in the literature on generalized linear models, this represents a specific relationship between the mean of a random variable and a linear combination of predictors, called the *link* function,

$$\log\left(\frac{y}{(1-y)}\right).$$

Research on the best link function is continuing at the Census Bureau, with possibilities that include logit, probit, loglog, and robit. An incorrect link function would result in poor extrapolations to situations that do not occur in the P- or E-sample data, unnecessary interaction terms in the model, and other typical results of lack of fit. The panel suggests that if the Census Bureau uses the Hosmer-Lemeshow goodness-of-fit test, it may help to choose the appropriate link function: that test will indicate whether an alternative link function would provide a better fit to the data.

Several complications would remain to be addressed.

Software for Alternate Link Functions. If it is discovered that an alternate link function is preferred, it might require a modest amount of software development to implement. However, this should be relatively straightforward in either SAS or R, which are two standard statistical software systems that the Census Bureau uses.

Loss Function or Objective Functions for Assessing Fit of Models. Another complication is that the current loss function underlying the fitting of the coefficients of these logistic regression models is implicit in the separate likelihood equations for the two models and is therefore somewhat disconnected from the ultimate goal, which is to predict the population size or, what amounts to the same thing, net coverage error. It may be that the ultimate goal can be better represented by weighting the likelihood equations to take this modified objective function into account. The Census Bureau has done some work in this direction and we support this research and its implementation if it is found to provide preferred estimates.

Measurement Error. Census data are subject to measurement error, and these errors will have deleterious effects on the application of logistic regression models. If the measurement error is unrelated to the outcome (match status or correct enumeration status), the effect on the data is the attenuation of relationships. In other words, the predictors will not be as effective without the measurement error. But if the measurement error is related to the outcomes, the effect could be much more complicated, including the introduction of severe biases.

Appendix C

Biographical Sketches of Panel Members and Staff

Robert M. Bell (*Chair*) is a member of the Statistics Research Department at AT&T Labs-Research. Previously, he worked at the RAND Corporation, where he directed a number of studies on social policy issues. His research interests include analysis of data from complex samples, record linkage methods, and machine learning methods. He is a member of the board of trustees of the National Institute of Statistical Sciences. He is a fellow of the American Statistical Association, chair of its Committee on Fellows, and has served on its Census Advisory Committee. He is currently a member of the National Research Council's Committee on National Statistics and has served on several of its panels. He has a B.S. degree in mathematics from Harvey Mudd College, an M.S. degree in statistics from the University of Chicago, and a Ph.D. in statistics from Stanford University.

Lawrence D. Brown is the Miers Busch professor in the Department of Statistics at the Wharton School at the University of Pennsylvania. His general areas of research include statistical decision theory, nonparametric function estimation, foundations of statistics, including admissibility and complete class theory, sampling theory (census data), and empirical queuing science. He is currently serving on the U.S. National Committee on Mathematics, and he is a fellow and past president of the Institute of Mathematical Statistics. He is a fellow of the American Statistical Association and the recipient of its Wilks Memorial Award. He is a member of the National Academy of Sciences, and he has served on several committees and panels of the National Research Council. He received a B.S. degree from the California Institute of Technology and a Ph.D. degree from Cornell University.

Michael L. Cohen is a senior program officer for the Committee on National Statistics, currently serving as study director for this panel and the Panel on the Design of the 2010 Census Program of Evaluations and Experiments. Formerly, he was a mathematical statistician at the Energy Information Administration, an assistant professor in the School of Public Affairs at the University of Maryland, and a visiting lecturer in statistics at Princeton University. His general area of research is in the use of statistics in public policy, with particular interest in census undercount, model validation, and robust estimation. He is a fellow of the American Statistical Association. He received a B.S. degree in mathematics from the University of Michigan and M.S. and Ph.D. degrees in statistics from Stanford University.

Roderick Little is the Richard Remington Collegiate professor of biostatistics at the University of Michigan, where he also holds appointments in the Department of Statistics and the Institute for Social Research. He was previously a professor in the Department of Biomathematics in the School of Medicine at the University of California at Los Angeles, an American Statistical Association/Census Bureau/National Science Foundation research fellow at the Census Bureau, and a scientific associate at the World Fertility Survey. His research interests include statistical methods for missing data and survey research methodology. He has served as coordinating and applications editor of the *Journal of the American Statistical Association*. He is a fellow of the American Statistical Association and has received its Wilks Memorial Award. He is a national associate of the National Academy of Sciences and has served on many committees and panels of the National Research Council. He received a B.A. degree in mathematics from Cambridge University and M.Sc. and Ph.D degrees in statistics from Imperial College of Science and Technology of London University.

Xiao-Li Meng is chair of and a professor in the Department of Statistics at Harvard University. Previously, he was a professor of statistics at the University of Chicago and a faculty research associate at the National Opinion Research Center. He has served as editor of *Bayesian Analysis,* as an associate editor of *Biometrika*, the *Journal of the American Statistical Association* and the *Annals of Statistics*, and is co-editor of *Statistica Sinica.* He is a recipient of the 2001 award from the Committee of Presidents of Statistical Associations for outstanding statistician under the age of 40. He is a fellow of the Institute of Mathematical Statistics and of the American Statistical Association. He is an expert on Bayesian methods and methods for missing data. He received a B.S. degree in mathematics from Fudan University in Shanghai, China, and M.A. and Ph.D. degrees in statistics from Harvard University.

Jeffrey Passel is a senior research associate at the Pew Hispanic Center. He is an expert on immigration to the United States and the demography of racial and ethnic groups. He formerly served as principal research associate at the Labor, Human Services and Population Center of the Urban Center. Prior to that, he was a demographer at the U.S. Census Bureau, including service as assistant division chief for population estimates, and he received the Bronze Medal from the U.S. Department of Commerce. He is a fellow of the American Association for the Advancement of Science and of the American Statistical Association. He is a recipient of the Demographic Diamond as one of the five demographers/social scientists selected by *American Demographics* as the most representative of influential work in the last 25 years. He has authored numerous studies on immigrant populations in America, focusing on such topics as undocumented immigration, the economic and fiscal impact of the foreign born, and the impact of welfare reform on immigrant populations. He received a B.S. degree in mathematics from the Massachusetts Institute of Technology, an M.A. degree in sociology from the University of Texas, and a Ph.D. degree in social relations from Johns Hopkins University.

Donald Ylvisaker is emeritus professor of statistics at the University of California at Los Angeles, having previously been on the faculties of Columbia University, New York University, and the University of Washington. His primary research interest is in the design of experiments; his applied interests have developed as a consulting statistician, frequently on legal matters. In 1990–1991 he held a joint statistical agreement with the Census Bureau under which he reviewed the Census Bureau's 1986 test of adjustment-related operations in Los Angeles, and he also served on the Advisory Panel to the Committee on Adjustment of Postcensal Estimates in 1992. He was an associate editor of a special issue of the *Journal of the American Statistical Association* on census methods. He is a fellow of the American Statistical Association and of the Institute of Mathematial Statistics. He received a Ph.D. degree in statistics from Stanford University.

Alan Zaslavsky is a professor in the Department of Health Care Policy (Statistics) at Harvard Medical School. His methodological interests include surveys, census methodology, microsimulation models, missing data, hierarchical modeling, small-area estimation, and applied Bayesian methodology. His health services research focuses primarily on developing methodology for quality measurement of health plans and understanding their implications. Another area in which he has contributed is methodology for measuring racial and ethnic disparities in health care delivery and determining their causes. He has written extensively on issues concerning the decennial census, including weighting and administrative records. He has served on the Census Advisory Committee on Adjustment of Postcensal Estimates and he has served on several panels of the National Research Council. He is a fellow of the American Statistical Association. He received an M.S. degree in mathematics from Northeastern University and a Ph.D. degree in applied mathematics from the Massachusetts Institute of Technology.

COMMITTEE ON NATIONAL STATISTICS

The Committee on National Statistics (CNSTAT) was established in 1972 at the National Academies to improve the statistical methods and information on which public policy decisions are based. The committee carries out studies, workshops, and other activities to foster better measures and fuller understanding of the economy, the environment, public health, crime education, immigration, poverty, welfare, and other public policy issues. It also evaluates ongoing statistical programs and tracks the statistical policy and coordinating activities of the federal government, serving a unique role at the intersection of statistics and public policy. The committee's work is supported by a consortium of federal agencies through a National Science Foundation grant.